PRAISE FOR ELIOT COLEMAN

"The incomparable Eliot Coleman . . . serious, meticulous, inspiring."

—*New York Times*

"One of America's most innovative farmers."

—Michael Pollan

"Eliot Coleman's book will help market gardeners establish the vital and profitable link between farm and city. Every small-scale grower and serious gardener should have a copy."

—Robert Rodale

"[Coleman] writes not just about gardening but about everything that connects to good food and pleasure; a Renaissance man for a new generation."

—Dan Barber, chef, Blue Hill and Blue Hill at Stone Barns

"I know of no other person . . . who can produce better results on the land with an economy of effort and means than Eliot. He has transformed gardening from a task, to a craft, and finally to what Stewart Brand would call 'local science.'"

—Paul Hawken, author of *Blessed Unrest*; editor of *Drawdown*

"Anybody seriously tempted to try . . . raising healthful food on healthy land . . . must first read *The New Organic Grower*. Coleman, who has been a quiet leader in the American organic movement for several decades, presents a balanced, logical exposition of his subject."

—*Horticulture*

"Every page is imbued with the wisdom and careful observations [Coleman] and his associates have gathered; from soil structure to 'mobile greenhouses' . . . each method is thought through to its ultimate impact on the earth and on economic survival."

—*Library Journal*

"This is the best book on small-scale farming I've read in years."

—Pat Stone, *Mother Earth News*

"From first sentence to last, [*The New Organic Grower*] is a delight—an earnest guide written with an impish sense of humor. It will refresh anyone who wants to get the most from a vegetable garden yet doesn't want to devote too much time and energy to the process."

—*Publishers Weekly*

"Coleman conveys a vast amount of detailed information without ever insulting the intelligence of the reader. He speaks as if to a fellow home or market gardener, sharing what works for him and discussing what he knows and what he doesn't know. *The New Organic Grower* will be the book you dog-ear and feather with yellow sticky pages, returning to it time and again."

—*San Francisco Chronicle*

The

NEW ORGANIC GROWER

The

NEW ORGANIC GROWER

A Master's Manual of Tools and Techniques for the
Home and Market Gardener

30TH ANNIVERSARY EDITION

ELIOT COLEMAN

Photographs by Barbara Damrosch

Chelsea Green Publishing
White River Junction, Vermont
London, UK

Project Manager: Patricia Stone
Editor: Makenna Goodman
Copy Editor: Laura Jorstad
Proofreader: Eileen M. Clawson
Indexer: Linda Hallinger
Designer: Melissa Jacobson

Printed in the United States of America.
First printing September, 2018.
10 9 8 7 6 5 4 3 2 1   18 19 20 21 22

Our Commitment to Green Publishing

Chelsea Green sees publishing as a tool for cultural change and ecological stewardship. We strive to align our book manufacturing practices with our editorial mission and to reduce the impact of our business enterprise in the environment. We print our books and catalogs on chlorine-free recycled paper, using vegetable-based inks whenever possible. This book may cost slightly more because it was printed on paper that contains recycled fiber, and we hope you'll agree that it's worth it. Chelsea Green is a member of the Green Press Initiative (www.greenpressinitiative.org), a nonprofit coalition of publishers, manufacturers, and authors working to protect the world's endangered forests and conserve natural resources. *The New Organic Grower* was printed on paper supplied by LSC Communications that contains at least 10% postconsumer recycled fiber.

Library of Congress Cataloging-in-Publication Data
Names: Coleman, Eliot, 1938– author.
Title: The new organic grower : a master's manual of tools and techniques for the home and market gardener /
  Eliot Coleman ; photographs by Barbara Damrosch.
Description: 30th anniversary edition [Third edition]. | White River Junction, Vermont : Chelsea Green Publishing, 2018.
  | Includes bibliographical references and index.
Identifiers: LCCN 2018015247| ISBN 9781603588171 (pbk.) | ISBN 9781603588188 (ebook)
Subjects: LCSH: Vegetable gardening. | Organic gardening. | Truck farming. | Organic farming.
Classification: LCC SB324.3 .C65 2018 | DDC 635—dc23
LC record available at https://lccn.loc.gov/2018015247

Chelsea Green Publishing
85 North Main Street, Suite 120
White River Junction, VT 05001
(802) 295-6300
www.chelseagreen.com

To Scott and Helen Nearing with thanks

One of the intangible legacies the Shakers left to the world is their demonstration that it is possible to create the environment and the way of life we want, if we want it enough. We can choose.

The Shakers were practical idealists. They did not dream vaguely of conditions they would like to see realized; they went to work to make these conditions an actuality. They wasted no time in raging against competitive society, or in complaining bitterly that they had no power to change it; instead they built a domain of their own, where they could arrange their lives to their liking.

—Marguerite Fellows Melcher,
The Shaker Adventure

CONTENTS

PREFACE TO THE
30TH ANNIVERSARY EDITION

Small farms are where agricultural advances are nurtured. New ideas are conceived every day by the folks who are solving Nature's puzzles. Since I turned in the original manuscript for this book years ago, I have traveled to Europe many times to see what was new there; spent time with organic growers on two trips to Australia, during a month in Chile and Argentina, and on a week in Mexico; and continued to develop and refine my thinking. I have benefited from the suggestions of those who read the book and wanted more information on certain subjects or wanted data on areas I did not cover. New equipment options are now being imported or manufactured here in the United States. And in some cases I have finally made up my mind, by acquiring more information, on points where I was ambivalent.

The revisions run the gamut from small changes in detail, to adding a lot of new material to some chapters, to adding whole new chapters where appropriate. I have added scientific references in endnotes for readers who wish to pursue a particular subject in greater depth. I have expanded the material in the chapters on plant-pest balance because the subject is of particular interest to me and of great significance in understanding how the organic vegetable grower fits into the natural world.

I have a rule that I write only about those things I know how to do. Consequently, there are a number of topics that this book doesn't cover. When I think there are techniques outside my own experience that readers might find useful, I refer them to someone who practices those techniques.

Throughout this book's creation and revision, my principal desire has been that it prove helpful to growers and gardeners. To that end I am pleased to have found nothing to change because it was wrong, but rather just areas that I could make clearer or on which I needed to expand because our practices have developed considerably. When I finished writing the first edition of this book in 1988, I said to myself, "I wish I'd had a copy 20 years ago." It is rewarding to be able to pass along the dependable information I have learned from other farmers around the world, dug out of obscure sources in many libraries, and devised on my own. There are coherent patterns in natural biological systems that can be adapted to producing the highest quality vegetables. I hope this book helps make those patterns more accessible.

CHAPTER ONE

The Other Side of the Tapestry: An Agricultural Metaphor

Still it seems to me that translation from one language into another . . . is like looking at Flemish tapestries on the wrong side; for though the figures are visible, they are full of threads that make them indistinct, and they do not show with the smoothness and brightness of the right side.

—MIGUEL DE CERVANTES, *Don Quixote,*
part 2, chapter 62

Imagine, if you will, an enormous Flemish tapestry hanging from the ceiling of a grand hall. The tapestry depicts the natural world in all its elegance; subsoil and topsoil, plowed fields and green pastures, gardens and orchards, prairies and forests, valleys and mountains, sea and sky are all crisply represented. There are creatures large and small, birds and fishes, bacteria and fungi, predator and prey, and the dynamic balances between them.

From where you stand on the front side of that tapestry, you don't notice too many others with you. There is, surprisingly, a great buzz of noise coming from the other side. When you walk way down to the far end of the hall and peer around the corner, you

can then see the tapestry's reverse side. With its stray colors and loose threads, it gives, as Cervantes suggests, only a vague picture of what is truly represented. But what you find there are enormous crowds of people actively trying to decipher what they see and trying to solve problems that only exist on the backside of the tapestry. They have no idea that there is a front side, and when you mention it you can tell they don't believe you. From where they stand, they can't see the elegance of the natural world. The vagueness of the backside has convinced them that the planet is poorly designed for food production and needs a great deal of help from humankind to straighten it out.

The problem isn't that the people on the backside are ignorant. On the contrary, many of them are brilliant. Their leading scientific disciplines—such as Discordant Thread Theory and Random Color Hypothesis—are highly respected and extensively researched. The university Department of Untrimmed Ends enrolls many student applicants, eager to make careers in the field. A multitude of learned disquisitions are published in numerous scholarly journals. Huge industrial complexes have arisen in concert with their line of thinking, and countless tons of stimulating and controlling substances are produced every year. The backsiders are convinced that as long as they keep expending enormous effort to compensate for Nature's flaws, all will be well.

However, when you step back to the front side of the tapestry, there are no flaws to be seen. You wonder if the backside people prove ecologist Frank Egler's statement, "Nature is not more complicated than we think. Nature is more complicated than we can think." But that is obviously not the case on the front side. As you study the front side more thoroughly, you begin to see the patterns involved. You notice that the agricultural practices of the frontside farmers are designed to harmonize with the directions in which the natural world wants to go anyway. You notice how those practices have been selected to nurture and enhance the systems with which they interact. Biology is in charge in the natural world, and a sensible agriculture is a biological agriculture that focuses on the soil—a biological agriculture that will continue to be productive as long as the earth abides. To make that a reality, we just need to move the discussion around to the front side of the tapestry.

CHAPTER TWO

A Little History to Begin

When knowledge is not in order, the more we have of it, the greater will be our confusion.

—HERBERT SPENCER

During the early decades of the 20th century, a group of concerned farmers in Germany, France, and England began trying to create a purely "biological," as opposed to chemical, agriculture. They had come to the conclusion that the use of chemicals in farming (both soluble fertilizers and toxic pesticides) was harmful—not only to the health of the soil, but also to the health of the livestock and the humans eating the produce from that soil. Their ideas were the inspiration for today's organic agriculture. This new (or I might better say rediscovered) biological focus for agriculture had arisen from their personal experience with the negative effects of chemicals on the quality of staple foods. These "biological farmers" were suggesting that agriculture was on the wrong path.

The first hints of this alternative agricultural thinking arose in the 19th century. An increasing number of farmers, scientists, and rural philosophers had questioned the wisdom of the chemically based inputs for food production that had grown quite prominent following their birth in the 1840s. They were supported in this quest for a more natural agriculture by advances in the biological sciences during the late 19th century, such as new discoveries that explained the workings of nitrogen fixation, mycorrhizal association, soil microbiology, allelopa-thy, weed ecology, and plant-pest relationships. The new biological sciences allowed for a deeper understanding of natural processes and showcased the intuitive brilliance of age-old practices like crop rotations, green manures, mixed stocking, and compost making. They also laid the scientific foundation for how a modern biologically based agriculture might be formulated.

The first organized objections to the use of agricultural chemicals came from the health-oriented Life Reform movement in Germany. Beginning in the 1890s there was increasing discussion about the decline in food quality that some suspected was due to excessive chemical fertilization. Even more objectionable were the toxic residues of the lead and arsenic pesticides used so liberally in those days. In response, the Life Reform movement stressed the benefits of food grown on fertile soil with natural methods. Gustav Simons's 1911 book, *Bodendüngung, pflanzenwachstum, menschengesundheit* (Soil fertility, plant growth, human health) was one of the earliest books expressing the growing sense of direct connections among the quality of the soil, the quality of the plants grown, and the quality of the resulting human nutrition.[1]

If you look back at the first flush of organic farming's notoriety in the United States during the 1940s, the names most often mentioned, the Englishman Sir Albert Howard and the American J. I. Rodale, rather than being the initiators, were just popularizers of that groundswell of new ideas that had begun to develop some 50 years earlier. It is important today, in light of the unceasing propaganda that tries to make organic farming sound like a pipe dream, to emphasize that it has a long history that is both practical and scientific.

The focus of the new biological approach can be best expressed in the three questions the early biological farmers hoped to answer:

1. How can long-lasting soil fertility be achieved and maintained?
2. How can the nutritional value of food crops be optimized?
3. How can pest problems in agriculture be prevented?

As we shall see below, the answers to all three of them are closely intertwined.

Soil Fertility

An English physician, G. Vivian Poore, included in his 1893 book *Essays on Rural Hygiene* a chapter titled "The Living Earth." Poore celebrated the indispensable role of organic matter as the basis of soil fertility, and the biological activity of the soil creatures as the power behind it. "Farmers and market gardeners will tell you that chemical fertilizers have 'got no bottom in them,' that their use is, so to say, a speculation.... With organic refuse, however, the case is entirely different, and the effect of the application of organic matter . . . to the soil is plainly discernable for three or four years . . . until finally . . . these organic matters become fertile 'humus,' which is the only *permanent* source of wealth in any country."[2]

Even more emphatic is Robert Elliot, an English farmer whose 1898 book, *Agricultural Changes*, was very influential in laying the groundwork for a biologically based agriculture. Elliot deplored the way the chemical companies tried to substitute chemical fertilizers for good farming. "Let me now briefly enumerate the other effects of humus. It not only supplies nitrogen, but as it decomposes, renders available some of the phosphorus and potash of the soil . . . Air, moisture, and warmth, which are all so necessary for the germination of seeds and the growth of plants, are but little influenced by the chemical constituents of the soil, being all more dependent on its physical condition, which can only be effectively influenced by large quantities of humus . . . the chemist really knows nothing of agriculture . . . for otherwise he would first of all inquire whether the farmer does make a full use of all the natural resources at his disposal. . . ."

In 1910 an American agricultural scientist, Cyril Hopkins, who was at that time director of the Illinois Agricultural Experiment Station, published his best-known book, *Soil Fertility and Permanent Agriculture*.[3] (A very up-to-date-sounding title today when we are looking for sustainable systems.) Hopkins's research led him to the following conclusion: "For all of the normal soils of the United States . . . there are only three constituents that must be supplied in order to adopt systems of farming that, if continued, will increase, or at least permanently maintain, the productive power of the soil. These are limestone, phosphorus, and organic matter. . . . The supply of organic matter must be renewed to provide nitrogen from its decomposition and to make available the potassium and other essential elements contained in the soil in abundance, as well as to liberate phosphorus from the raw material phosphate naturally contained in or applied to the soil."[4]

These three examples emphasize that from the earliest days of chemical fertilizers, both farmers and researchers saw clearly their flaws. Dismayingly, the economic power and political influence of the chemical companies allowed them to continue pushing their products. Nonetheless, the new biological soil-fertility concepts were eagerly investigated during the early 20th century in Europe and gained an increasing number of adherents. By the 1930s the concerned farmers had begun formulating an updated, scientifically grounded version of the biologically based agriculture that had preceded the chemical/industrial invasion. The efforts of Heinrich Krantz and Ewald Könemann in Germany,[5] Hans and Maria Müller in Switzerland,[6] Pierre Delbet[7] and Raoul Lemaire[8] in France, and Lady Eve Balfour[9] and Sir Albert Howard[10] in England all led to the development of successful farming systems based on continuous improvement of soil fertility using natural methods.

By 1946 these ideas were being discussed with sufficient interest among agriculturists in general that the International Harvester Company, a major American producer of agricultural equipment, sponsored the publication of a 125-page pamphlet authored by Karl Mickey, *Health from the Ground Up*.[11] It was a sequel to an earlier soil conservation volume, *Man and the Soil*, also by Mickey, but this second publication dealt "primarily with the influence of soil characteristics on the individual." Mickey made very positive comments about the work of Weston Price, Sir Robert McCarrison, and Sir Albert Howard, names known then, almost exclusively, in the alternative agricultural world. When discussing fertilizers he wrote, "It is not uncommon for the addition of a nitrogen fertilizer to a soil low in other nutrients to cause abnormal growth and disease in the plants . . . excessive or unwise use of fertilizers containing of pure chemical salts may hasten the depletion of some of the vital secondary elements in the soil. These are conserved by the use of stable manure, the plowing under of legumes, and other methods of replenishing the organic content of the soil."

Nutritional Quality

The second theme in this discussion, the effect of the fertility of the soil and the growing conditions on the resulting nutritional quality of the produce, has a less scientifically respectable but nonetheless long and enthusiastic parentage. The obscure names of Charles Northen, Julius Hensel, Albert Carter Savage, Sampson Morgan, Royal Lee, and Weston Price all surface in the course of a thorough search of old sources. Their major concern was that if the full complement of minerals was not present in the soil (both major minerals such as calcium and potassium and trace elements like copper, zinc, and boron), the nutritional quality of the food and the subsequent health of the consumer would be diminished. The general suggestions for improving soil mineral availability ranged from raising the percentage of organic matter in order to increase biological activity, to growing deep-rooting green-manure crops that would bring minerals up from lower soil layers, to supplying natural soil minerals by spreading the finely ground rock dust that was generally available as a waste product of the rock-crushing industry. A number of reasonably respectable reviews of natural mineral sources were written in response to growing interest, but it was the actual awareness of mineral deficiencies affecting people and livestock that finally brought attention to the issue.

Dr. E. C. Auchter, then chief of the Bureau of Plant Industry of the United States Department of Agriculture (USDA), wrote a lead article for *Science* magazine in 1939 titled "The Interrelation of Soils and Plant, Animal and Human Nutrition." He stated that up to then we had neglected "the interrelationship between the physical well being of man and the factors in the soil that affect the composition and development of plants." Whereas the previous focus had been only on large yields, "We ought to give more attention to producing crops of the highest *nutritional* quality for man and animals . . . if conditions of lowered health exist in part because of low quality plant or animal products produced on deficient soils, then the plant, animal and soils investigators have a challenge and responsibility that can not be shirked." In a subsequent article he concluded, "The unfolding of this relationship may conceivably revolutionize agricultural theory and practice and profoundly change our ideas on the advancement of human welfare."[12]

Ten years later Paul Sears of Yale University, in an address to the Ohio State Medical Association, mentioned his thoughts on the subject. He discussed groups representing the "growing popular interest in the relation of Soil to Health. . . . The main tenet of such groups is that, by promoting the normal biological processes in the soil and returning all possible organic matter to the soil, healthy plants, livestock, and human beings will be produced. At times some of the more enthusiastic over-do things. . . . But they are, in general, on the track of a very important truth."

The Role of Pests

Starting at almost the same time as the movement advocating naturally based soil fertility, a parallel group of researchers were reevaluating the general attitude toward pests in agriculture and coming to very different conclusions than the pesticide enthusiasts. According to this alternative line of thinking, pests were not enemies to be killed but rather indicators to be heeded. Pests, they contended, could not gain a foothold unless the plant had been negatively predisposed by inadequate growing conditions. They stressed that the solution was to improve the growing conditions. This idea had been around for quite a while. Even as far back as 1793, in a letter to his daughter, Thomas Jefferson wrote: "I suspect that the insects which have harassed you have been encouraged by the feebleness of your plants and that has been produced by the lean state of the soil."[13]

Erasmus Darwin, Charles's grandfather, speculated in 1800 that the leaves of a fruit tree damaged by insects were "previously out of health, which occasioned them to supply a proper situation for those insects which molest them."[14] Thomas Green Fessenden, author of a garden book, *The New American Gardener*, that was enormously popular in the 1830s, stated: "The preventive operations are those of the best culture . . . choice of seed or plant, soil, situation, and climate. If these are carefully attended to, it will seldom happen that any species of insect will effect serious and permanent injury. Vegetables which are vigorous and thrifty are not apt to be injured by worms, flies, bugs, etc."[15]

In 1870 Vincent Gressent, a market gardener in Paris, wrote *Le potager moderne*, an instruction book for Parisian market gardeners: "For vegetable growing chemical fertilizers don't do all that one wants; they stimulate the plant and produce quantity, but to the detriment of quality. . . . In principle, insect pests only attack weak, sickly plant specimens lacking proper nutrition. . . . In proof of this I offer the market gardens of Paris where vegetable growing has reached perfection. . . . One does not see pest problems in Parisian market gardens wherever copious compost use and rational crop rotations are practiced by the growers" [my translation].[16]

Back in the early 19th century, the appearance of fungal and bacterial plant diseases was explained by the Autogenic Theory. According to the autogenicists the effect of environmental factors on the plant was the prime cause of plant disease; the visible symptoms were exterior manifestations of interior malfunctions. That theory had been swept aside by the end of the 19th century as agriculture acknowledged the existence of the bacteria and fungi associated with plant diseases following the work of Pasteur, Koch, and others. As a consequence, consideration of the role of environmental factors as root causes was relegated to a minor role. There were, however, a number of investigators who, although acknowledging the new pathogenic theories about the function of microorganisms, still credited the influence of growing conditions. They contended that microorganisms could only incite disease when the host plant had previously been rendered susceptible by unfavorable environmental conditions (poor soil, excess moisture, insufficient air, low temperature, and so on), which the alert grower had every possibility of controlling.

H. Marshall Ward (1854–1906) in England and Paul Sorauer (1839–1916) in France were two of the leading proponents of these ideas, which came to be

known as Predisposition Theory. *Predisposition* was defined as the tendency of non-genetic conditions (such as those mentioned in the parenthesis above), acting before infection, to determine the susceptibility of plants to pests. This is a varying degree of resistance or susceptibility dependent on external causes. A major cause is the quality of the soil. In his essay "The Nature of Disease" (1905), Sorauer wrote that "for the production of a parasitic disease the presence of the parasite alone is not determinative but the constitution of the host organism is also a determining factor.... That condition of a living creature which we are accustomed to term 'healthy,' without being able as yet to define it, is one such restricting limit which the parasite under normal conditions is not able to overcome."[17] Ward in his Croonian Lecture of 1890 effectively summed up the predispositionists' case. "Disease is the outcome of a want of balance in the struggle for existence just as truly as normal life is the result of a different poising of the factors of existence."[18] It is a tribute to Ward's perceptive thinking on this subject that 85 years

later, Baker and Cook in their *Biological Control of Plant Pathogens* stated the same idea in almost identical terms: "The occurrence of a plant disease thus indicates that some aspect of the biological balance is not in equilibrium."[19]

In 1938 Von H. Thiem, a German researcher, discussed the existence of both an absolute and a relative immunity to pests. The former he called genetic immunity and the latter, pheno-immunity. He considered plants to be genetically immune when their resistance is such that a specific pest will never propagate and develop on them. Pheno-immune plants, on the other hand, are those whose degree of resistance is influenced by outside factors. If the resistance of a pheno-immune plant is to be maintained, then cultural conditions such as soil type, fertilization, moisture, and so forth must be carefully considered. Thiem even contended that monoculture, long considered a causative factor in insect multiplication, would present no problem if proper cultural practices succeeded in assuring the pheno-resistance of the crop.[20]

H. Marshall Ward's influence reentered the natural farming discussion in 1940 with the publication of Sir Albert Howard's book *An Agricultural Testament*. It turns out that when Howard was engaged in graduate work at Cambridge from 1896 to 1898, H. Marshall Ward was his major professor. Howard's wife acknowledged that Ward was the one who introduced Howard to the predisposition idea.[21] Howard incorporated the concept into his own pursuit of a natural agriculture and stated emphatically in *An Agricultural Testament* that "insects and fungi are not the real cause . . . but only attack unsuitable varieties or crops imperfectly grown. Their true role is that of censors for pointing out the crops that are improperly nourished."[22]

The introductory discussion above presents some background for the three strains of thought that coalesced into a new biologically based concept of agriculture, first called organic farming in the 1940s:

1. Soil fertility can be raised to the highest levels by techniques that increase the percentage of soil organic matter, by rotating crops and livestock, and by maintaining soil minerals through using natural inputs such as limestone and other finely ground rock powders.
2. The plant quality resulting from doing #1 correctly provides the most nutritious possible food for maintaining human beings and their animals in bounteous health.
3. The plant vigor resulting from doing #1 correctly stimulates the plants' immune systems and renders plants resistant to pests and diseases.

All three begin with and depend on how the soil is treated. But in the best small-farm tradition, the fertility of that crucial soil factor is not a function of purchased industrial products. It evolves from intelligent human interaction with the living processes of the earth itself. These are processes that are intrinsic to any soil maintained with organic matter. They are what the earth does.

Detractors have often misrepresented a biologically based agriculture as if it is nothing but the substitution of purchased organic inputs for purchased chemical inputs. Even if there were evidence to document the rationale for a substitution philosophy, it would lose on the grounds of economics alone. Both bonemeal and dried blood, for example, two popular "organic" fertilizers, are prohibitively expensive on a farm scale. Furthermore, such substitution thinking is not pertinent to the actual objective of a biological agriculture—namely, the development of sustainable, farm-generated systems for maintaining soil fertility. The concern is not the substitution of one fertilizer for another but rather the long-range practical and economic viability of farming practices. Supplies of bone- and blood meal are no more assured than are supplies of chemical fertilizers that derive from finite and dwindling resources. Agricultural systems that rely on inputs from either source cannot be depended on over the long term. That which can be depended on, however, is a biologically focused farming system that bases fertility maintenance on proven cultural practices with the addition of locally available waste products.

Among those cultural practices I include:

Crop rotation. Firmin Bear of Rutgers has stated that a well-planned crop rotation is worth 75 percent of everything else the farmer does.

Green manures. Deep-rooting legumes not only fix nitrogen, penetrate hardpan, and greatly increase soil aeration but also bring up new mineral supplies from the lower depths of the soil.

Compost making. Of all the support systems for the biological farm, none is more fortuitous than the world's best soil amendment, compost, which can be made for free on the farm from what grows thereabout.

Mixed farming. Raising animals and crops on the same farm has both symbiotic and practical benefits. The crop residues feed the animals and the animal manures feed the soil.

Multispecies cover crops. Using eight or more species maximizes yield of the next crop in the rotation due to enhanced soil health, extra biomass, ecosystem diversity, and better weed suppression, among other factors.

Ley farming. The fertility of land plowed up for row crops after two to four years (depending on soil type) in grass/clover pasture is practically that of virgin soil because of the enormous amount of plant fiber added by the perennial plant roots and the biological stimulation of the manure from the grazing livestock.

Undersowing. Establishing a green-manure crop underneath the growing cash crop can often double organic-matter production in the course of the year without any effect on the cash crop.

Rock minerals. The slow, measured release to plants of locally sourced mineral amendments (calcium, phosphorus, potassium, trace elements, et cetera) added to the soil as ground rock powders mimics that from natural soil particles. The nutrients are made available to the plant roots by the biological soil life nurtured by soil organic matter.

Enhancing biodiversity. This includes practices such as growing a wide range of crops, sowing pastures with many different forbs in addition to grasses and legumes, carrying a mixture of livestock, establishing hedgerows for beneficial insect habitat, setting up bird nesting boxes, building a pond for additional habitat, and so forth. The more components involved, the more stable the system.

These practices are pieces of a management program by which biological farmers successfully work through natural processes to meet human needs in food and fiber without in any way overwhelming those processes and causing them to malfunction. The best biological farmers follow a pattern at odds with the pattern of chemical agriculture. The chemical/industrial mind-set focuses on the symptom of a problem and devises expensive products in order to palliate that symptom. The biological/organic mind-set focuses on the cause of the problem and looks to aid natural processes in such a way as to correct the cause and prevent the problem from ever occurring.

Biological farmers use the age-old fertility-enhancing practices mentioned above to correct the cause of low soil fertility rather than attempting to treat the symptoms (poor yields, poor quality) by purchasing chemical stimulants. The same pattern applies to pest problems. By improving soil fertility, avoiding mineral imbalance, providing for adequate water drainage and airflow, and growing suitable varieties, biological farmers avoid the plant stress that causes pest problems, thus correcting the cause rather than treating the symptoms—insects and diseases—with pesticides. As Albert Howard wrote in 1946, "I have not hesitated to question the soundness of present-day agricultural teaching and research—due to failure to realize that the problems of the farm and garden are biological rather than chemical. It follows, therefore, that the foundations on which the artificial fertilizer and poison spray industries are based are also unsound." What it boils down to is that the long-term goal of a biological agriculture is to cultivate *ease* and *order* rather than battle futilely against *disease* and *disorder*.

But can we really farm that way? Can a successful agriculture be conducted by simply combining the known effects of natural processes with the management provided by intelligent human understanding of how to nourish those processes? If that type of agriculture can work and could be made universal, then this new agriculture would be truly sustainable and have the power to transform the world. Back in 1967 when I began farming, none of us paid attention to whether the agricultural science of the moment (as opposed to agricultural tradition) approved of our approach. We started farming organically with compost and cultural practices because the ideas made sense, and lo and behold, they worked. Alternative agricultural research today is showing that we were pretty astute. Studies are appearing almost too fast to read them all.

For example, the importance of soil organic matter is more appreciated every day even though it "is arguably the most complex and least understood component of soils."[23] Bioactive humic substances produced by earthworms in compost have been found to enhance root growth and availability of nutrients "by mechanisms that are not yet clear" because "relatively little attention has been paid."[24] Work with composts has determined their ability to control plant diseases through initiating in the plant what Harry Hoitink of Ohio State calls Systemic Acquired Resistance.[25] T. C. R. White has explained how the effect of stressful growing conditions "upsets the metabolism of the plant in such a way as to increase the availability of nitrogen in its tissues," which increases "survival and abundance of herbivores feeding on those tissues" even though "these physiological changes may often not be sufficient to produce visible signs of stress in the plant."[26] Non-genetic tolerance to stress is a form of "induced resistance dependent on environmental factors" and an approach "that only a limited number of researchers have tried to define."[27] But even genetic resistance makes no difference if negative growing conditions inhibit the expression of the genes. In USDA research to determine why tomatoes growing in mulch of vetch green manure were more disease-resistant and longer-lived than identical tomatoes with black plastic mulch, Kumar et al. found that the genes for longevity and resistance were not "turning on" in the sections without the vetch mulch.[28]

Nutritionists to their dismay have found what they call "dilution effects" in Green Revolution–type crops. Breeding programs aimed to produce high-yielding cultivars combined with intensive chemical fertilization to push yields still higher have resulted in vegetable and grain crops that do not have their full nutritional complement because of the inability of their limited root systems to absorb enough of the

minor nutrients. The result for human beings is a "hidden hunger" caused by trace element deficiencies in those who consume such foods. The recent study by Brian Halweil, *Still No Free Lunch*,[29] presents a very complete picture of the relationships among plant breeding, high chemical fertilizer use, the nutritional quality of the resultant produce, and the new interest in pursuing such research. Other forward-thinking scientists around the world are beginning to investigate biological issues, and they are finding that the biologically focused system that organic farmers have been creating for the past 125 years, this alternate agricultural reality, is as good as the farmers have claimed it to be.

So here is the interesting question. How could these ideas have been so obvious, so logically presented, and yet so consistently ignored by the majority of agricultural scientists? I can imagine three simple explanations for the failure of agricultural scientists to comprehend the existence of a different reality, for why they cannot imagine a world where soil preparation using compost and green manures and rock minerals creates high yields of vigorous plants that do not need the protection of pesticides and fungicides. There seems to be great difficulty in comprehending what I call a *plant-positive* approach (strengthening the plant through optimal growing conditions to prevent pests) as opposed to the conventional *pest-negative* approach (killing the pests that prey on weak plants). As Benjamin Walsh stated back in 1866 in *The Practical Entomologist*, "Let a man profess to have discovered some new Patent Powder Pimperlimplimp, a single pinch of which being thrown into each corner of a field will kill every bug throughout its whole extent, and people will listen to him with attention and respect. But tell them of any simple common-sense plan, based upon correct scientific principles, to check and keep within reasonable bounds the insect foes of the farmer, and they will laugh you to scorn."[30]

The first explanation is the lack of a word. There is no word in our popular vocabulary to describe *plant-positive* thinking. We all know what the Department of Plant *Pathology* (*pathos*—suffering) concerns itself with. But does any university have the antonymic Department of Plant _____? What would the word be? Sanology (from the Latin *san*—health) or Euology (from the Greek *eu*—good) might be suggested as possible new words. Or possibly call it the Department of Plant Phylactotrophy? (*Phylact*—protect; *troph*—nourish.) What if all the land-grant schools had a Department of Eucrasiotrophic Agriculture? (*Eu*—good; *crasio*—constitution; *trophic*—nourishing.) What if we lived in a world where we had the expectation of healthy plants rather than pest-ridden plants? What if the Department of Phytostenics (*phyto*—plant; *sten*—strength) published research explaining how plant health had to be subverted through mistaken cultural practices before pests could dominate? That would be a different world. But the fact remains that it is difficult for most people to comprehend a concept so novel that their language has never had scientific words to define it.

The second explanation is that humans cannot imagine a world where they are not in charge. As a biological farmer, I work in partnership with Nature, and I'm a very junior partner. Given the limited amount of hard knowledge available, I often refer to my management style as "competent ignorance," and I find that a very apt description. But my level of trust in the design of the natural world and willingness to be guided by it is discomforting to those who think we should exercise total power over Nature. Thomas Colwell, in his chapter in the book *Human Values and Natural Science*, is most emphatic on this point: "But though part of Nature, man's unique function . . . lies in controlling and transforming the natural world, not piously seeking its guidance. How profoundly we believe this today. How could we help but believe it; the entire edifice of our civilization is built upon it. The Baconian conception of science as control over nature is not only an intellectual presupposition of ours, it is a deeply implanted emotional attitude as well."[31]

The third explanation goes back to the beginning of the industrial revolution when the money world began to replace barter and exchange. At that point what would have been seen as the great benefit of a biological production system—minimal need for purchased inputs—suddenly came to be seen as its defect. In an industrially dominated money economy, the processes by which biological agriculture produces food are downright subversive. Because they are self-resourced through that partnership with the natural world noted above, they are independent of industry. By "self-resourced" I mean that for those participating in a biological agriculture, the majority of the inputs are coming from within the farm. Thus, biological farmers who take full advantage of the earth's contributions do not need to purchase industry's products.

That may explain why so few people are aware of the simple ways by which perceptive farmers have learned to successfully satisfy human needs for food and fiber within the framework of Nature's biological realities. By being self-resourced, the techniques of a biological agriculture offer no foothold for industry, resulting in no advertising, no research and development, no buzz, no audience, no business. If everyone can grow bounteous yields of vigorous plants that are free of pests by using homemade compost and age-old biological techniques, there is no market for fungicides or pesticides or anhydrous ammonia. If a concept cannot be commodified—which is to say, if it isn't dependent on the purchase of industrial products—industry is antagonistic, and the idea gets short shrift in our commercially dominated economy.

But maybe the problem is that we just don't believe any of this is possible. What? Farmers can grow broccoli without green worms? Livestock can be raised without antibiotics? Dream on! But I have come to these conclusions and can suggest these radical ideas because of what I see happening on my farm every day. We often jokingly refer to our farm as the National Empirical Research Station. When scientific evidence is lacking, practical experience is all we have to go on. And the facts are right in front of my eyes while I am cultivating or transplanting or tilling or mending fences. I see that the biologically based agriculture I have practiced for the past 50 years really works. When I have done my job as a farmer correctly; when I have optimized the biology of crop production by maintaining soil organic matter, improving soil aeration and mineral balance, and providing adequate moisture; when I have paid close attention to enhancing natural processes, there is no downside. The livestock are in full health. There are no green worms on the broccoli. There are no root maggots in the onions. The yield and the quality of my farm products are consistently exceptional without any need for industrial products. That is the daily reality of a successful biological farm. Could it be that the people have been conned into ignoring the brilliance of a whole other way of farming by a limited worldview that has never allowed them to consider non-commodifiable options?

The reality of today's world is that the practical success of the many farms managed on biological lines coexists with the striking lack of interest (antagonism actually) from scientific agriculture in exploring why these farms succeed. The foundation on which our Maine farm operates—a sense that the systems of the natural world offer elegantly designed patterns worth following—appears to be an indecipherable foreign language to most of agricultural science.

This book is suggesting that the techniques espoused by chemical agriculture are based on the mistaken premise that the natural world is inadequate and, therefore, needs to be supplemented with industrial chemicals. Because that misconception creates further problems, chemical agriculture requires enormous patch-up artifice in order to function. Pesticides, fungicides, miticides, and so forth are all parts of the patch-up system. This is a familiar tale. True believers in chemical agriculture, especially where there is enormous economic incentive, are reluctant to admit that their initial premise was mistaken, and they cling to it in the face of obvious evidence to the contrary.

CHAPTER THREE

Agricultural Craftsmanship

Why and how do plants grow? Why and how do they fail? Why do plants seem to grow successfully for some people and in some places and not others? The answers lie in those factors that affect the growth of plants, including light, moisture, temperature, soil fertility, mineral balance, biotic life, weeds, pests, seeds, labor, planning, and skill. The grower can influence some of these factors more than others. The more they can be arranged to the crop's liking, however, the more successful the grower's operation will be.

The Biology of Agriculture

Working with living creatures, both plant and animal, is what makes agriculture different from any other enterprise. Even though a product results, the process is anything but industrial. It is biological. We

You should have both field and greenhouse crops.

are dealing with a vital, living system rather than an inert manufacturing process. The skills required to manage a biological system are similar to those of the conductor of an orchestra. The musicians are all very good at what they do individually. The role of the conductor is not to play each instrument, but rather to nurture the union of the disparate parts. The conductor coordinates each musician's effort with those of all the others and combines them in a harmonious whole.

Agriculture cannot be an industrial process any more than music can be. It must be understood differently from stamping this metal into that shape or mixing these chemicals and reagents to create that compound. The major workers—the soil microorganisms, the fungi, the mineral particles, the sun, the air, the water—are all parts of a system, and it is not just the employment of any one of them, but the coordination of the whole that achieves success.

I remember a conversation I had back in 1979 with a Kansas farmer in his 60s who farmed some 700 acres (2.8 square kilometers). His methods were considered unconventional at the time because he had always farmed without purchasing herbicides or pesticides and bought only small quantities of lime and phosphorus. I asked him on what theory he based his farming. He said there really wasn't any theory that he knew of. It was simply the same now as it had ever been. He mentioned a favorite book of his, a 1930s agricultural textbook that stressed the value of biological techniques such as crop rotation, animal manures, green manures, cover crops, mixed cropping, mixed stocking, legumes, crop residues, and more. He said he used those practices on his farm simply because they worked so well. The book never mentioned any "theory" and probably never knew one. The book referred to these biological techniques as "good farming practices."

My Kansas friend assured me that by basing his crop production on those good farming practices, he enjoyed yields equal to and often far better than his neighbors'. He'd seen no yield increase from soluble fertilizer when he had tried it. His crop-rotation and mixed-farming system made weeds, pests, and diseases negligible problems. When fertilizer prices rose he felt as secure as ever because his production techniques were so fundamentally independent of purchased materials. And as long as those good farming practices worked and continued to make his farm profitable, he would continue to use them. He concluded by saying that, if there were any theory involved, he would simply call it "successful farming."

I have long followed similar good farming practices—biological techniques—on my farm. The secret to success in agriculture is to remove the limiting factors to plant growth. These practices do that efficiently and economically by generating a balanced soil fertility from within the farm rather than importing it from without. They power the system through nurturing the natural processes of soil fertility, plant growth, and pest management, enabling them to work even better When chosen carefully and managed perceptively so as to take full advantage of specific aspects of the natural world, these good farming practices are all the farmer needs. As a further bonus they eliminate such problems as soil erosion, fertilizer runoff, and pesticide pollution at the same time.

Creating a System

I have been compiling and evaluating information on biologically based food-production techniques ever since I started farming. At first I collected this material as a commercial vegetable grower because I needed it to ensure the success of my own operation. In the process I became aware of the enormous untapped potential of this way of farming and enthralled by the discovery and practice of the simple techniques of an agriculture that is in harmony with the natural world.

In order to develop a dependable vegetable-production model, I concentrated on collecting information in four subject areas:

1. How to simplify production techniques
2. How to locate the most efficient machinery and tools
3. How to reduce expenditures on purchased supplies
4. How to market produce in the most remunerative manner

From my experience, these four areas represent the basic information needed for small-scale, economically successful, biologically based food production.

The first category explains just how straightforward and rational a successful vegetable-production system can be. Although growing commercial crops is often considered for "experts" only, it most emphatically is not. The world of plants is vital, vigorous, and self-starting. Drop a seed in the ground, and it wants to grow. The common wisdom possessed by successful farmers is that they understand how to help the seed do what it is already determined to do. The more successful the farmer, the better the understanding of how to enhance the natural processes without overwhelming them. That simply stated idea is the key to successful organic food production.

Next is the importance of efficient and dependable machinery and tools that match the needs of small-scale production. Small farmers can and do compete and succeed economically and practically when they have access to equipment scaled and priced within their means and designed for their specific tasks. The fact that such useful and appropriate equipment was not readily available has been a contributing factor in the demise of the small farm and the concurrent belief that it cannot succeed. All too often, unwarranted and problematic growth in farm size has been dictated by the need to justify expensive and oversized equipment because nothing else was available.

In order to find, try out, and modify the right equipment, I have looked all over the world. The equipment ideas included in this book originated in many different countries. The recommended tools do their jobs admirably. New models will no doubt appear in the future and should be even better. But I expect the basic relationship of the tasks to the system to remain fairly constant.

Third, the economic success of any operation must be ensured. In order to keep costs down, I emphasize the importance of "low-input production practices." By that I mean practices such as crop rotation, green manures, animal manure management, efficient labor, season extension, and so forth. Production benefits are gained from careful management rather than expensive purchases. Not only will these practices save money in the short run, but they will also increase the stability and independence of the farm in the long run. The more production needs are farm-generated or labor-saving, the more independent and secure the operation becomes. The farm and its economy cannot then be held hostage by the unavailability or high prices of commodities from outside suppliers. The most stable farm economy is one that is built on the greatest use of farm-generated production aids.

Finally, no matter how successful I might be in the first three areas, it would be of little use to me if I did not have a successful marketing program. Marketing has always been the make-or-break area for small-scale producers. Much depends on highly developed marketing skills that probably would not have led someone to farming in the first place. The recent growth of farmers markets has in many instances helped the marketing of local produce. But there are other solutions. I have noticed on both sides of the Atlantic that farmers who enjoy the greatest economic success have found competitive niches in the larger marketing system. The extent of this market for small-scale growers and ways to reach it are described in the chapter on marketing.

Learning How

There are a number of ways to learn any new skill. You can jump in boldly right from the start and count on the sink-or-swim reaction to carry you through.

You can work as an apprentice to someone else who knows how to do it. Or you can go to college for a course of study. I followed the first method, and I recommend it to anyone who enjoys that type of challenge. You learn quickly because you have to. But be forewarned that it can be an occasionally stressful and exhausting adventure—especially if you start with minimal resources, as we did, and have to practically create your world before you can inhabit it.

For many people the second option is preferable. The management pressure is on the employer and you can concentrate on details. I think it has a much higher success rate than the third. You will have a better chance to succeed if you have learned well by working for a good grower than if you have studied agriculture in college. The apprentice systems of past centuries turned out very competent practitioners of their crafts. When you want to learn how to do something, go straight to those who are doing it well. In addition to the invaluable hands-on experience you'll gain, you'll also become more motivated for any supplementary reading and book-learning (remember how unmotivated many of us were in school?) because you will have had some solid background and your questions will arise directly from your own experience and interest.

In one of my favorite books, *The Farming Ladder*, author George Henderson relates both his experience as an apprentice working for other farmers as well as his later role as a master farmer taking pupils of his own.[1] During his learning period Henderson worked on four different farms, chosen to give him as broad a background as possible. Most impressive to me is the way he threw himself fully into the job. He never missed an opportunity to work harder or longer whenever it appeared there might be something new to learn. He tells a story of when he was battling a fire in a haystack—coughing from the smoke, covered with the soot, soaked by the water— and a neighboring farmer who happened by told him he was a fortunate lad because he was getting all this experience at someone else's expense.

Later, when he himself was a successful farmer, Henderson treated his pupils as he had treated himself. He provided good training and expected hard work in return. He speaks about compensation systems that were common in those days. The student received room and board but was expected to pay the farmer a monthly fee for the first three months. After three months, if the student was a competent worker, the farmer returned that same fee every month as salary for the second three months. That way, if the student was shiftless and soon departed, the farmer would be paid for his trouble. However, the student who persevered and worked hard would receive six months of thorough training at no cost. Given the price of a college education, the system sounds pretty good. An eager student who wants to work with an exceptionally gifted farmer might want to make a similar proposal in order to be taken on.

After discussing his experience working on other farms, Henderson distills the wisdom he learned into a couple of sentences. It matches closely with my own experience as a self-taught beginner. "Good farming is the cumulative effect of making the best possible use of land, labor, and capital. It is not the acreage you farm, but the intensity of production you maintain, which determines the financial success of the venture."

But It Can't Be Done, Can It?

Most sections of the United States were once fed by small local farms. Today that is considered an impossible dream. Even where professional farmers are involved, the idea of the economically viable small farm is criticized as visionary. Many agricultural experts state that it just cannot work. Their opinion is based primarily on economic and production conclusions drawn from large-scale agricultural operations. Unfortunately, little consideration has been given to the advantages inherent to the small end of the spectrum.

If you understand how the economic and practical realities change when low-cost production methods are allied with the right machinery and marketing practices, then the case does not seem hopeless at all. In fact, the negative opinion of the "experts" is contradicted by the number of successful examples of small-scale food-production operations both here and abroad. Those numbers are increasing daily as improved low-cost technolo-

1968: carving a farm out of the wilderness.

An Important Question

Although I intend this book to be comprehensive, there is one question that readers must answer for themselves. Until the answer to this question is resolved, all the best instruction in the world won't help. Let me address you directly. The question is, "Why do you want to be a farmer?"

I suggest sitting down with pen and paper and coming up with some answers to that question. I've found that the best way to sort out fuzzy thinking is to compose ideas in a readable form. Is it only the idealized lifestyle that you crave? A touch of "over the river and through the woods to grandmother's house we go"? Think carefully about whether a desire to farm is a positive action toward farming or a negative reaction against what you do now.

Dissatisfaction with your present career, an intolerance for city living, or a perceived lack of excitement in your life may generate that negative reaction. Future hopes are often naively focused on rural life because it fulfills the bucolic fantasies that we all share in the back of our minds. I encourage thinking long and hard about what you really want to change, where you really want to go, and why.

In a negative reaction, the would-be farmer has suddenly had enough of the city, the dead-end job, or simple boredom and jumps impulsively for the fantasy of fresh air and farm living. In contrast, a positive action stems from a long-term desire to farm, which may have been set aside for practical or economic reasons. Such an action is based on knowledge of the hard work and discipline that a career in agriculture demands. In a positive action, the move has been planned carefully, and the farmer is only waiting for the parts to fall into place and the time to be right.

Whether the motivation is understood or not, there is one option that clearly makes sense and is worth repeating—go and work on a farm. Try out the idea by laboring (and learning) with someone else. Experience the good times and the bad, the realities and the rewards. If possible, work on more than one farm. The more background and experience you can get, the better off you'll be.

The requirements for success in farming are like those for any small business: organizational aptitude, diligence, financial planning, the ability to work long hours, and the desire to succeed. Added to these are the need for skill at working with your hands, sensitivity toward living creatures, a high level of health and fitness, and a love of what you are doing. Farming offers a satisfying challenge found in no other profession to those who can meet its demands, overcome its difficulties, and reap its rewards.

gies become more widely available and consumer demand grows for high-quality local produce.

From my experience, an area as small as 1 to 2 acres (4,000 to 8,025 square meters) can offer a highly productive scale for vegetable growing. The management skills needed for an operation that size are enjoyable rather than onerous. It is a comprehensible size for commercial food production—large enough to make a living, yet small enough to retain the emphasis on quality; diverse enough so that the work is never dull, yet compact enough so it is never out of control.

There is a distance, to be sure, between the isolated example and the consistent success. Consistent success can only result if the system makes practical sense, has been well tested and proven over a number of years, and is followed with diligence and understanding. The experts have been mistaken before, and they will certainly be mistaken again. What they have failed to realize in the case of the small farm is that, with careful planning, organization, and desire, there is nothing that "can't be done."[2]

CHAPTER FOUR

Land

Everyone shares a kinship with the land. No matter where we are in time or distance, the desire for an ideal country spot is very real. Whether the image comes from books, childhood experiences, or the depths of our souls, it has an indelible quality. The dream farm has fields here, an orchard there, a brook, and large trees near the perfect house, with the barns and outbuildings set off just so. The dream is effortless. The difficulty comes in trying to find such a place when you decide to buy one.

I suggest not trying to find that perfect place. Rather than the finished painting, look for the bare canvas. Every ideal farm at one time began as field and woodland. Its transformation was the result of some predecessor's planning, organization, building, and management. This is a satisfying process in itself, and the end result may be far more successful if it springs from the changes the farmers make themselves. This is not a hard-and-fast rule, merely a suggestion. If you already own or can obtain a

Adding organic matter will make any soil more fertile.

Four Season Farm photographed from a drone.

productive farm that is well established, then by all means do so. But if that is not possible, then do not hesitate to buy the raw land and create the farm.

A few suggestions follow on things to take into account when looking at a piece of land with an eye toward turning it into a successful small-scale farm.

Soil Type

Almost any soil can be made productive for growing crops. The difference lies in the amount of effort needed to make it so. The less ideal a soil is to begin with, the more attention must be paid to modifying its characteristics. Extreme soil types require an inordinate amount of work. Pure clay or pure sand and gravel are obviously less desirable than a rich loam. On the other hand, while the transformation of imperfect soils requires time and energy, the result can be as productive as initially

more promising soils. I have grown magnificent produce on a very sandy-gravelly soil that began with a pH of 4.3. It took a few years of manuring and liming, adding phosphorus, potassium, and trace elements, growing green manures, and establishing a rotation, but it became, and has remained, a highly productive piece of ground.

The dream soil for vegetable production is sandy loam. This is a general term describing the proportions of three ingredients: clay, silt, and sand. Clay consists of fine particles that help the soil hold water and provide a potentially rich storehouse of plant nutrients. Sand consists of larger particles, mostly silicates, that keep the soil open for air and water penetration and aid early warming in the spring. Silt falls somewhere between these two. A fourth soil ingredient—humus or organic matter—is the key to productivity. It opens up heavy clay soils for better air and water movement and easier working. Humus also helps hold together and give structure to light sandy soils, creating more stable conditions for the provision of water and nutrients to plants.

My advice is to look for the best soil possible but not to be put off if it is not perfect. The cultural practices recommended in this book will help put in the organic-matter and nutrient supply to make a productive success from a wide range of initial soil types.

Soil Depth

Old adages apply to soils as well as to people. Don't be fooled by a pretty face. What is underneath the surface of the soil will be important in the future, so investigate it from the start. Investigations with a spade or shovel can be augmented by requests to the local Natural Resources Conservation Service office for information on that particular soil or others like it in the area.

Soil depth should be considered in three ways, because some soil problems are more easily modified than others. First consider the depth to bedrock. This can't be changed to any degree. So ask hard questions about it and take cores with a soil sampling auger if you are not satisfied. Rock outcrops in the surrounding topography are warning signs of a shallow soil. Second, there is the depth to the water table. Land that otherwise appears acceptable may have a seasonally high water table during wet times of the year. This could make early-spring planting difficult or impossible. A too-high water table also limits the depth of usable soil by hindering root penetration. In most cases the condition can be cured through surface or subsurface drainage, but these modifications are expensive.

The third consideration is the depth of the topsoil. There are layers in the soil that make up the soil profile. The uppermost layer, the topsoil, is commonly from 4 to 12 inches (10 to 30 centimeters) deep. If you dig a hole, it is relatively easy to determine where the darker topsoil ends and the lighter subsoil begins. Normally, the deeper the topsoil the

better, since topsoil depth is closely related to soil productivity. This is the one of the three soil-depth factors that can be most easily modified over time. Subsoil tillage, manuring, deep-rotary tillage, and the growing of deep-rooted, soil-improving crops will all help to deepen the topsoil. These techniques will be described in later chapters.

Aspect and Slope

The lay of the land in most of the northern half of the United States is a very important factor. Land with a southern aspect has a number of advantages. A southern exposure warms up sooner in the spring. (Of course, the reverse is true in the Southern Hemisphere.) The more perpendicular a slope is to the angle of the sun, the faster it warms up. The grower who plans to compete for the early vegetable market needs this advantage. Land in the Northern Hemisphere at about 43 degrees latitude (the northern border of Massachusetts, Illinois, or California) that slopes 5 degrees to the south is actually in the same solar climate as level land 300 miles to the south.

The logical application is to choose land with a southern slope for early crops.[1] A southwestern slope is preferable to a southern or southeastern slope. On a southwestern slope, less direct radiation is required to evaporate dew or frost in the early morning, and more is available for absorption once the initial daily warming has occurred. However, in this as in other areas, a compromise may have to be made.

Air Drainage

Plants and gardens need to breathe fresh air. A low area with no air movement is undesirable for a number of reasons. Stagnant air encourages fungal diseases, holds air pollutants around plants, and stays colder on frosty nights. The last is an important consideration, because cold air, which is heavier than warm air, "flows" downslope and collects in hollows and valleys. A farm in such a location is at a great disadvantage. Conversely, one on a hillside or near the top of a slope is better protected from frosts because the cold air of late spring and early fall flows down the hillside and settles in the valley. In many cases this cold-air drainage can result in a longer frost-free growing season of two and even three weeks at both ends of the season.

Of course you don't want the land to be so sloped that it is either difficult to farm or subject to erosion. Actually, a flat area on the edge of a slope enjoys as much air drainage as the top of the slope does and is probably the ideal situation.

Wind Protection and Sunshine

Although some air motion is good, more is not always better. An excessively windy site can cause physical damage to plants, cause windborne soil erosion, and provide less-than-ideal growing conditions due to the cooling action of the moving air. A windbreak, which can be anything from trees and tall hedges to low stone or board walls to strips of wheat or rye, can minimize the damaging effects of strong winds and optimize the benefits of solar warmth. The temperature of the soil is raised in shelter because, as more surface warming takes place, the accumulated heat is conducted downward into the soil. Windbreaks also help to create more ideal growing conditions by preventing the loss of transpired moisture. Reduction in evaporation means a consequent reduction in heat loss. In many ways a wind shelter does more than just lower wind speeds. It creates a beneficial microclimate far different from that in nearby ground.

Unfortunately, too much of a good thing always causes a loss somewhere else. Too much wind shelter can mean inadequate sunshine. A balance must be struck somewhere between adequate wind protection and excessive shade. In order to achieve that ideal in the right place and at the right time, you must pay close attention to the path of the sun. It is not enough merely to say that the sun is higher in

the sky in summer than in winter. It also traces a different path in the sky, rising and setting farther to the north in summer and to the south in winter. The ideal May windbreak should be planned so as not to shade the January greenhouse.

Sunlight is the motive power for photosynthesis. Every effort must be made to take full advantage of it, especially during the early and late months of the growing season. Sometimes that means cutting down a tree to reduce shade. When faced with this situation, I think long and hard about whether the shade cast is serious enough to warrant removal of the tree. If the decision goes against the tree, I don't hesitate to cut it down, but I always replace it with another one in a more suitable location. Sunshine is one of the most dependable free inputs in a food-production system, and the grower should make every effort to maximize its contribution.

Water

Excess water is a flood and too little water is a drought. The farmer must make provision for both. River-bottom land is usually fertile and easy to work, but the grower must have access to higher land for dependable early crops in wet years. If the flood-prone land can be planned for a summer rotation or later crops, it will be less likely to cause a failure. Droughts are the other side of the coin. Water supply for irrigation is a major concern in some areas of the country. On average, provision should be made for applying 1 inch (2.5 centimeters) of water per week during the growing season if it is not supplied by natural rainfall. When looking for a good piece of land for vegetable growing, water is a key factor.

The ideal solution is a year-round spring or stream that can be tapped into and that is sufficiently higher than the cultivated fields to allow for gravity feed. The piping for such a system can be laid inexpensively on top of the ground in spring and drained in fall, thus avoiding the expense of burying a long water line. Second best to a gravity system is a dependable pond from which water can be pumped. A well is a good third choice, but it is difficult to be sure of an adequate supply to meet the needs of production at the height of the season. The choice of whether irrigation water will be applied to the fields by surface flow, subirrigation, overhead sprinkling, or drip systems will depend on topography, climate, and quantity of water available.[2] On this farm I have always preferred overhead sprinkler irrigation. I find it much easier to work with than all the complications of drip. However, your best source of advice is the experience of other local growers in your area.

Geographic Location

This is too broad a subject to cover in detail. North, south, east, or west; hot or cold; wet or dry; urban, suburban, rural, or remote: Crops can be grown in any and all of these places. The main consideration regarding location is proximity to market. Obviously, a suburban location close to many customers, stores, and restaurants is appealing, but the cost of the land in such areas is high. Land is less expensive in rural areas, but a greater effort must be made to develop and reach an adequate market. One compromise is to locate your farm in a rural area that attracts summer visitors. That market will not be available year-round, but it will appear during your most productive cropping season.

Access

Both you and your customers need to be able to get to and about your land. Access in both instances is often not taken into account. A market location on a busy highway must provide plenty of space for safe parking. If it doesn't, few will stop. A long, unpaved, or seasonally impassable road to the farm will also discourage customers and increase the costs of access and operation. Road access for deliveries should be large enough for tractor trailers. Streams on the property may be picturesque, but if they must be crossed often to get from one field to another, they become a liability. Land for a successful food-production operation should be easy to enter, work in, and work around. It should be accessible to large equipment required for deep cultivation and manure spreading. Careful consideration should be given to the flow of activities from seeding through harvest, and provision made so that access to one part of the operation is not hindered by activities in another.

Security

The depredations of both two- and four-legged invaders can prove costly to a small farm. Dirt bikes and other off-road vehicles can destroy a whole spring's work in a thoughtless moment. Fields of raspberries, strawberries, and other fruit are tempting targets for midnight thievery unless they are protected by a fence or are located near inhabited dwellings.

Crows, pigeons, and other winged invaders of newly seeded crops can be stopped in a couple of ways: either by beating them or by joining them. To beat them the grower might use the floating crop covers described in chapter 24. These covers protect seedlings from the birds. But joining the birds can also work. An old-timer once gave me the obvious answer to hungry crows: Feed them. His method was to place small piles of corn in easily accessible places around the field, figuring that crows with

The gate for the deer fence at the entrance to our farm.

full bellies won't pull up newly sprouted corn. The feeding continues until the corn is past the susceptible stage. I have followed his suggestion for many years and am very fond of this ingenious and benign solution.

Four-footed marauders also need to be considered. In my part of the world deer can graze off a whole field of newly transplanted seedlings in a few hours. Raccoons are notoriously fond of sweet corn crops. The mere mention of woodchucks, pigeons, gophers, and the like can put an evil gleam in the eyes of many growers. I know the same situation exists with rabbits, kangaroos, llamas, and elephants in other regions of the world. Although there are a number of temporary measures (chemical repellents, organic repellents, radios tuned to all-night talk shows, and scarecrows, to name a few), the most consistently successful barrier is a good fence.

One effective fencing material is the lightweight plastic mesh with ¾-inch (2 centimeter) squares that was originally sold as bird netting. Now it is also advertised as deer fence because it can be wrapped around shrubs to prevent deer from browsing in winter. I purchase it in 100-foot (30 meter) lengths that are 7½ feet (2.2 meters) wide. In trials where I merely draped the material around the edge rows of stands of sweet corn, it proved 100 percent effective against raccoons for a few nights. I have also kept deer out of a large area by attaching the mesh to posts (I used 2×2s) driven in at a 12-foot (3.7 meter) spacing. This material deters the critters not by strength but by trickery. It feels like a large and frightening spider's web. I can understand how the wild invaders must react by comparing my discomfort when it has caught in the buttons of my shirt while I was installing it. If the material is rolled up and put away when not in use, it will last a long time.

A heavier-weight black plastic mesh of the same height but with larger openings is sold as an "invisible" deer fence. The installation instructions suggest running it through the woods just back from the edge of the cleared area. It is supported by attaching to tree trunks with fence staples. The black color cannot be easily seen in the shady woods—it is almost invisible. It is also very effective. I erected it through the woods on one side of a field to shut off a very active deer access route. It forced the deer to either end of the fence, where they were further deterred by bramble hedges and human habitation. I have also used this material successfully (supported by temporary 1-inch (2.5 centimeter) PVC electrical conduit posts woven through the mesh) as the only animal protection around temporary trial plots. No varmints got in.

Another "invisible" deer deterrent concept involving dogs was researched by the Missouri Department of Conservation.[3] They used the same electronic containment fence and shock collar that suburbanites use to keep pets in their yards. The containment "fence" in these systems is a single wire that can be buried, attached to existing fences, or laid on top of the ground. The dog wears a receiving collar and is trained to the shock it will administer if the dog tries to cross the fence. This idea ingeniously solves two problems of using dogs for deer protection. If the dog is tethered, it loses its effectiveness after a short while, because the deer become habituated to the barking. Untethered dogs, on the other hand, will run deer past the boundaries of the field, thus breaking game laws. The containment fence and collar allow the control advantage of an untethered dog while limiting it to the area you need protected.

The Missouri study suggests that breeds like Australian shepherds, blue heelers, and border collies will be most effective because they have a patrolling and chasing instinct. Two dogs are better than one so they have company; neutering and spaying are recommended. The dogs should live in the field and not alternate between field and farmhouse. The researchers also found the system works best if a 30-foot- (9 meter) wide, mowed buffer strip is left between the fence and the edge of the crop so the dogs can make a perimeter trail and mark scent. Further, placing doghouses, feeders, and waterers where traditional animal trails enter the field increases the chances of the dogs interacting with invaders.

The most secure, albeit the most expensive, option is an 8-foot- (2.4 meter) tall woven wire deer fence. Less expensive but almost as secure would be one of the high-tensile electric fence models. They can be designed to keep out everything from raccoons to deer, including even kangaroos and elephants, I suspect.

Pollutants

Pollutants are a fact of life in most areas. It makes sense to give some thought to this problem in order to know what can be done before it surfaces. Both lead from exhaust and cadmium from tire wear have been found in excessive amounts on food grown within 200 to 300 feet (60 to 90 meters) of heavily traveled highways. A 6- to 8-foot- (1.8 to 2.4 meter) high evergreen hedge bordering the highway can block out the bulk of this pollution before it reaches the fields beyond. Lead from old paint and plaster can be a serious soil pollutant in areas where buildings once stood. Under certain conditions lead and other heavy metals in the soil can be taken up by plants in amounts that are highly detrimental to consumers, especially young children.

Residues from toxic waste dumps can travel downstream or downwind at far greater distances than is commonly recognized. Crops irrigated with water from an aquifer contaminated by one of these sites will cause toxic residues to accumulate both in and on the crops. Areas downwind of large industrial smokestacks suffer a continual dusting of questionable substances, including gases such as ozone, which is directly harmful to plant growth.

Land that has grown commercial fruit or vegetable crops for many years may contain undesirable residues. Old-time agricultural poisons containing lead, arsenic, and mercury were used heavily on orchards and some vegetable crops in the first half of the 20th century and are known to have perma-nently compromised the future agricultural use of large areas of land. It is common practice nowadays to test the water quality in the well before buying a rural home. Similarly, it would also be wise to test the soil for heavy metals before purchasing old agricultural land.

These problems may exist to a greater or lesser degree. They are mentioned here as one more factor to be considered in selecting land on which to farm. As with anything that may affect the success of a farm operation, it is best to know beforehand. Nothing could be more discouraging than to put a lot of work into establishing a small farm and to then find that, because of unanticipated circumstances, the site will not produce quality food. Forewarned is forearmed.

Acreage

How much land should you purchase? Some people buy twice what is needed and sell half of it later to partially cover their costs. Others purchase more than they need initially so they can modify their plans in the future. Perhaps in addition to fields, you'll need wooded land for a firewood supply. You might want to purchase an adjacent area to set aside as wildlife or native plant habitats. Maybe what it comes down to is you just want some extra space.

All these decisions are personal ones. The premise of this book is that you can make a good living on 2 acres (8,100 square meters) or less of intensive vegetable production. Thus it is those acres that concern us most. If a few acres is all the land you can afford, it is more than sufficient for an economically successful farm. If you can afford more, another few acres perhaps, then more options open up. The crop rotation could include small livestock on pasture to take advantage of the long-term soil-fertility benefits of leaving land in deep-rooting grass and legume pastures for a few years before rotating it back into vegetable crops. (See chapter 9.) Another 1 or 2 acres might be used for berry crops. Beyond that, you must watch out for the trap of more, more, and more. It is easy to farm a lot of land with a pencil and paper, but a lot harder to actually do it. The best advice is to buy only as much land as you will use and to buy the best land you can afford.

Soil Tests

In my experience, the utility of soil tests is controversial. Soil tests appear to be part science and part necromancy. A sample of soil is sent to a laboratory where it is supposedly analyzed, investigated, and divided into its component parts. You then receive a report on its good and bad points. Despite the guise of the laboratory, there is more art than science to a good soil "test." The secret of using the results of a soil test lies in their interpretation. Not everyone agrees on how that should be done. Nor do people agree on how many useful conclusions can be drawn from the data. I recommend in all cases that you make an initial basic soil test on the fields you intend to use for vegetable production. Better yet, have two or three tests made on the same soil sample, each by a different laboratory. Have a state university lab make one of the tests and then send the same sample to one or two private labs. The results are often similar to the old saying, "A man with one watch knows what time it is; a man with two watches is never sure." Still, given the lack of certainty inherent in soil testing, it is better to have two or three opinions from which to draw your conclusions than only one.

The interpretation of soil test results is something that I suggest growers do for themselves. It won't be done mystically or according to one particular theory, but rather by common sense. If the test shows the soil is low in a nutrient, then add it. The method and manner of adding these nutrients is explained in chapter 12. My recommendation is to add nutrients in their most available but least soluble form. By "most available" I mean that they should be finely ground, and by "least soluble" I

mean not immediately water-soluble but rather gradually water-soluble through the action of soil bacteria and dilute soil acids—in other words, through the natural soil processes.

In taking a soil test you want to obtain as homogeneous a picture of the field or plot to be cultivated as possible. Instead of taking one sample from a spot in the middle of a field, a number of random samples should be taken over the whole area and then combined and mixed. This way the soil sent to be analyzed will be a more accurate representation of what you have. It is best to do this sampling and mixing with tools whose composition will not affect the results of the test. For example, if you are testing for iron, don't sample with a rusty iron trowel. In general, a stainless-steel soil probe for the samples and a hard plastic bucket for mixing them are best.

The Ideal Small Farm

I doubt you will find the ideal small farm (I never have), but the best of luck to you anyhow. The listing below describes what I believe the perfect piece of land would look like. Please remember that none of this is carved in stone. The determined farmer can transform even the most unlikely site into a model farm by applying the basic techniques of soil building.

For Sale: 30 acres. 20 acres mixed hardwood forest, 10 acres pasture. Description: Reasonably flat (about 5 percent slope to south or southwest), set on the brow of a hill with good air drainage to the valley below. Protected on the north by higher land and dense forest. Excellent year-round gravity water supply. Soil on the cleared land is clay loam on one half and sandy loam on the other half. Topsoil is 12 inches deep, well drained, and highly fertile, with very few stones and a naturally neutral pH; presently growing a grass-legume sod.

CHAPTER FIVE

Scale and Capital

The economists are always telling us that American farms have to get bigger to survive. Small vegetable farms are considered an anachronism. But the weakness of small farms has not been one of scale. It has been one of useless information. Like a traveler with an outdated map, the small farmer has had to rely on outmoded concepts. Since the standard information sources (government departments of agriculture, the agricultural colleges, the farm-supply industry) have not thought small farming possible, they have neither been compiling nor dispensing information on how to do it. To understand this situation we need to understand our underlying attitudes toward farming in America.

Bigger Is Not Better

The most common attitude holds that bigger is better, and the admonition to "get big or get out" is frequently heard. There is a basic assumption in agriculture that a farmer with a few acres in mixed production is less successful and less advanced than the farmer who practices industrialized monoculture on a huge scale. In the recent history of American agriculture, to take one example, such an assumption may have been defendable. American farmers with ambition and entrepreneurial ability took advantage of the resources of capital, transportation, and the agricultural technology that all blossomed in the 20th century to expand their operations, buy more land, and become increasingly specialized.

This should not, however, be seen as the only or even the best way to farm. The economic and environmental problems of large-scale agriculture now make headlines every day. It seems the warning should be changed to "get big and get out."

The major obstacle I had to overcome when I started farming was a lack of models. Back in 1965 there were almost no commercially successful organic small farmers from whom I could get inspiration and with whom I could share ideas. My prototype of the economically viable small farm didn't even exist. But highly productive small vegetable farms had once existed, so I began with the assumption that if it could be done before, it could be done again. The know-how had to be available somewhere. At the start, old books and old-timers were my best sources of information and support. They were a good beginning, but they left me in the past. Nevertheless, I made some headway, but in the process I became aware of the reasons why small farming had died out. The product was excellent, but the process was exhausting. It was neither cost-effective nor efficient. But I wasn't ready to give up. I knew there had to be suitable techniques, traditions, and equipment somewhere that would bring this production process up to date.

The European Model

There are sections of the industrial world where the small farm continues to succeed. Western Europe is one such area. There, the small farmer is an institution that didn't die. Whether through tradition or stubbornness, European small farmers have persevered, and the tool and equipment producers and information resources have kept up with them.

Entrepreneurial farming in Europe led to better rather than bigger. In part because more land was not readily available, increased income lay in improving production on existing acreage rather than expanding. But a more important influence was the love of good food. The connection between the qual-

ity of the produce and the careful attention of the farmer to the land has always been recognized and practiced in Europe. As a consequence, the output from the average European small farm is the type of exquisite produce that always finds a ready market. Europe's small farmers have traditionally produced the bulk of the food eaten in Europe and made a good living for themselves in the process.

The European farmers provided the model that I knew must exist. They inspired my farm. I saw no

Tool care is always worth the effort.

reason that regionally based small-scale food production could not be just as successful in the United States. Unlike grains, fully ripened fruits and vegetables are highly perishable. They are particularly well suited for local markets. As more and more consumers become disenchanted with the products of industrial agriculture, the availability of fresh, carefully husbanded, farm-ripened food is a tradition we will have the incentive and the market to reclaim. It is a tradition that demands small farms.

In the process of finding the models, the philosophy, the tools, and the technologies of the small farmer, I also learned a great deal about scale and the advantages of staying small. I learned right from the start that one must totally reevaluate the basis for size in farming. The best way to do that is to wipe the slate clean and start thinking from scratch. All the when, where, what, why, and how questions need to be asked anew. How much production can one person or family handle? What kinds of equipment and techniques can be efficiently employed on the family farm? What helps the farm family do more with less?

The 2-Acre Answer

The answers are illuminating. Two and a half acres (1 hectare) is the optimal size because it is about as much land as a couple or small family can manage for growing, harvesting, and maintaining fertility. By "manage" I mean both practically and professionally—practically, by using the low-cost equipment and simple techniques that fit the tasks, and professionally, by making sure there are enough people per acre to stay on top of things. To effectively manage a farm operation, I believe there is a person-to-land-area ratio that cannot be exceeded. Producing quality food requires an investment of effort on the part of the grower that naturally limits the amount of land farmed. For diversified vegetable growing, I place that upper limit at somewhere around 2 acres (8,095 square meters) per couple. That ratio is small only in numbers. Two acres is

more than sufficient land to grow a year's worth of vegetables for 100 people. Anyone feeding that many folks can honestly consider themselves to be running a highly productive farm.

Tool Background

When I began as a small-acreage vegetable grower over 50 years ago, the tools available were mostly from the 19th century—Planet Jr. wheel hoes and seeders plus traditional hand hoes. I knew we needed to find more efficient tools for commercial small-farm production. The tools at the heart of our modern, low-cost production techniques came from many sources.

First were those that I saw on farms in Europe back in the 1970s: walking tractor tillers, soil blocks, the broadfork, flame weeders, caterpillar tunnels, and Swiss wheel hoes.

And those I found in European hardware stores: the three-section bed roller, pinpoint seeders, bed preparation rakes, stirrup hoes, and improved one-row seeders.

Next came those that have recently appeared on the market: SolaWrap greenhouse film and drill-powered greens harvesters.

And, finally, those we designed ourselves: the Tilther, collineal hoe, wire weeder, wire hoe, row markers for the bed preparation rake, Quick Hoops, and the six-row seeder.

A French agricultural supply company, Terrateck, is presently showing great imagination by modifying tractor implements to make hand-scale production even more efficient, including a manual mulch layer, finger weeders, wheel weeders, and new wheel hoe attachments.

And the tool department at Johnny's Selected Seeds is doing yeoman service in locating and making available most of the above.

We have definitely left the 19th century.

In order to fully understand the potential for practical and economic success on the small farm, it must be understood that scale is not necessarily limiting or static. Size is only one of the components of an economic operation, and growth is more than just a change in size. It is generally accepted that a business must grow and respond to change in order to survive. The same applies to the small farm. But there are countless ways in which a farm can expand: quality, variety, and service are a few examples. I feel strongly that growth and change in the direction of better will ensure the economic and agronomic survival of the small farm more assuredly than growth and change in the direction of bigger.

It also helps if one can learn to ignore the well-meaning advice of economists, because their understanding of scale is industrial rather than agricultural. Their advice does not apply to biological production. The truth is that what can be accomplished on a small scale in agriculture cannot be duplicated on a larger scale. The small farmer's aim is to produce a quality product for an appreciative clientele. The production of large quantities of red-, green-, or orange-colored cellulose for a mass market is not the same thing.

The most encouraging aspect of small-scale farming is that the capital requirements to start up are reasonable. In this and many other ways small scale in agriculture should be understood as a very positive factor.

Equipment

The basic high-quality equipment needed to manage a few acres of land in vegetable production can be purchased new very reasonably. If used models are available, that cost can be reduced even further. However, since much of the equipment I recommend has only recently become widely available, that option may not exist. This equipment consists of:

- Walking tractor/tiller
- Wheel hoes
- One-row and multi-row seeders
- Soil-block equipment
- Hoes and hand tools
- Carts and wheelbarrows

The single most important concept that keeps the capital investment at this very reasonable level (and simplifies the skills needed for operation and maintenance of the equipment) is the size of the equipment. Instead of a four-wheeled tractor with costly implements, I recommend a much simpler two-wheeled walking tractor/tiller for soil preparation along with hand-powered tools for seeding and cultivating. Reliance on these tools is not a concession to economics. Their outstanding performance and flexibility alone recommend them.

Additional capital investment for equipment other than the above may be required (or at least desired by the farmer) under many growing and marketing conditions. In this category I include greenhouses and irrigation.

If the whole cost were spent and paid back over five years (the depreciation life of the equipment) at 10 percent interest, the total cost per acre per year on 2 productive acres (8,095 square meters) would be very reasonable. If you compare that with the $100,000-per-acre gross income an efficient retail vegetable operation should realize, you can see that the economics of small-scale vegetable farming are promising. If the local market permits a degree of specialization in the higher-priced crops, the income figure can be raised substantially. The marketing concepts suggested in chapter 23 offer ways to help achieve those goals.

The small farmer operates in a unique situation. Definitions of the possible, the economic, the realistic, and the practical are completely changed. More than anything else, a lack of understanding of these definitions and a parallel lack of information on downscaled biological and mechanical technologies have added to the belief that human-scale, regionally based agriculture "can't be done." It can.

CHAPTER SIX

Labor

"You cannot get good help nowadays." "People don't work hard enough." "People don't care." "They want too much money." "They aren't dependable." Labor can be a problem. Many of these comments may be valid, others not, but all are worth noting. It is wise to make some serious choices ahead of time before you find yourself muttering those very sentiments.

Family Labor

My suggestions in this area are consistent with the food-production premise of this book—small, manageable, and efficient. The family is the best source of labor for the small-scale farm. So the most important recommendation is to set up an operation that is small, manageable, and efficient enough to be run

Some of the wonderful young people who we have hired.

mainly by family labor. Why? Because farming is hard work, and the rewards at the start are measured more in satisfaction and pride than in large salaries. The farm family will do the work because it is their dream. It is their canvas, and they are painting it the way they've always wanted it to look. Hired help who can involve themselves from the start on such an intense level of participation are not easy to find.

This production system is planned to make the most of family labor in the following ways:

- I have chosen equipment for ease and efficiency of use and repair.
- I recommend growing a broad range of crops to spread the work more evenly over a long season.
- I take a management-intensive approach for fertilization and pest control.
- I stress forethought and pre-planning to avoid panic.
- I propose imaginative marketing approaches to save time and energy.

Most important, this system is based upon a philosophy that aims at stability by establishing long-term, self-perpetuating, low-input systems of production as opposed to short-term, high-input systems.

Outside Labor

The best-laid plans don't always run true, and chances are the grower will sometimes need outside labor. When paid helpers are required, I have some suggestions that may prove useful.

If you find good employees, plan to keep them. Pay a fair wage, and investigate profit-sharing options and other rewards. One good worker familiar with your operation is worth three inexperienced workers. Be imaginative. What does the farm have to offer that will attract the ideal people? The usual pool of labor available for part-time farmwork has never been the best. But think further. For many people farming is exciting. Most everyone has a

farming urge hidden behind an urbanized facade. While most are still dreaming about it, your farm is a reality. So offer potential helpers not so much a job as rather a part-time outlet for their dream. It is surprising how many people share this dream but have not yet decided to pursue it. Offer that reward to those people.

Finding Willing Workers

The potential labor pool extends from the young to the old, from students to retirees. Homemakers whose children are now in school or college are often looking for a new challenge. Working on an organic farm can give them meaningful part-time work and a chance to turn their energy and competence into valuable assets. There are many such people who are reliable, intelligent, hardworking, interested, and motivated and would love the chance to share in someone else's dream. For them the rewards are only partially financial. Since work hours are often limited to evenings or early mornings (harvesting for market, say), the possibility of fitting farmwork into standard schedules is increased. Where to look for willing workers? Some of the following are good places to start:

- Retirement communities
- Supermarket bulletin boards (put up help-wanted signs and specify the benefits)
- Local colleges
- Food co-ops
- Garden clubs
- Condominium and apartment complexes

Be Efficient and Flexible

Be efficient. Maximize skills, minimize deficiencies. Labor should be hired to do what the boss does not do best or what the boss does not need to do. Ideally, the boss is going to be good at growing and marketing. Fine. Then hire help to harvest, wash, crate, and distribute. Whoever is best suited for a certain area of the operation should spend his or her time doing that as well as it can be done. Overall efficiency will be greater. Hire labor to complement rather than replace family skills.

Be flexible. Work out a solution for the particular labor needs of the moment. If the labor arrangement of the farm does not parallel that of modern agriculture, let it be of no concern. Many unique situations are successful. A farm may be next door to a large vegetarian community that will buy everything at a premium and help out to boot. Students from a nearby college may provide all the part-time labor on a work-study program. The farmer may have a dozen brothers and sisters living nearby who eagerly come and help out whenever they are needed. Ignore any claims that a farm only succeeds because of a special arrangement. Success simply means that a farmer is doing something right. Remember, too, that no matter how good a deal you may have at the moment, it should never be assumed to be permanent. Always have an alternative solution or two on hand.

Getting Quality Work

Jobs should be done correctly. The complement to labor is management. That is what the boss must do, and the quality of management determines to a large degree how well labor performs. Horticulture is a skilled profession, and there is a need to work quickly but precisely. Standards must be set. I was impressed in Europe to see how horticulture is respected and understood. The employees are professionals, are proud of their work, and take satisfaction in doing it well. In the past this may have been true elsewhere, but rarely anymore. The boss has to instill that spirit of professionalism.

The repercussions of slipshod garden work are cumulative. Rows planted crookedly in a moment of carelessness cannot be cultivated efficiently and will require hand-weeding for the entire growing season.

Weeds that are allowed to go to seed one year will increase the weed problem for the next seven years. The quality of each job will affect the efficiency of the entire operation. Poor work must not be tolerated.

Along with quality production goes excellence of skills. Set work standards and stick to them. Most people have never learned the necessary bodily coordination needed to work well with simple tools. This lack of training and the consequent awkwardness result in making a job much more difficult than it needs to be. Show your helpers how it's done. They should be taught to use garden tools just as carefully as they would be taught to play a musical instrument or speak a foreign language. Remember, any physical work is made easier by planning the job out beforehand, working at an efficient rhythm, and dividing the job into attainable pieces.

Different spacing for different crops.

Inspiring the Crew

An important facet of management is attitude. Management must care about labor's satisfaction. Many people will come to work because they are interested, so encourage their involvement. Explain not only what the job is but where it fits into the overall scheme of things and why it is important. If someone is starting in the middle of a process, take a moment to explain it fully so they can see both the beginning and the end. Not only will they be more interested when they understand the rationale for their efforts, but once they see the whole picture they will be able to suggest improvements in the system. Very often a beginner has seen things that I have missed because I was no longer looking at the job from a fresh perspective.

Finally, one last suggestion for dealing with outside labor. As I have said, the farm family is its own best labor force because they are motivated. Think for a moment—why? Because they love what they are doing; because it is creative and satisfies a creative urge; because farming is necessary and fulfilling work; because quality is important, and good growers take pride in producing a quality prod-uct. Whatever the reason, the bosses must convey a sense of that to outside workers. Don't hesitate to be inspirational and enthusiastic. If it is the magic of transforming a tiny seed into daily bread, then say so. If it is the joy of providing customers with truly nourishing food that they can trust, talk about it. Not everyone will share the same motivation, but enthusiasm is contagious. Spread it about.

Firing Workers

There are times when it will be necessary to fire someone. Do it nicely, but don't put it off. There is nothing more frustrating than making do with uninterested and unmotivated workers. One determined griper can ruin the experience for everyone else. Nip it in the bud quickly. If there are valid gripes, they should be dealt with fairly and openly. But beyond that, be firm. Some people seem to enjoy complaining. I prefer not to have them around.

A farm can't do without labor. The trick is to do well with it. If outside labor can't be counted on, don't set up a system that relies on it. When outside labor is necessary, use the natural advantages of the farm to attract people who want to be there.

CHAPTER SEVEN

Marketing Strategy

Establishing a marketing strategy is one of the first steps in a successful vegetable operation. Which crops? How much to grow? Ready when? Sold to whom? The key to success in marketing is to pay careful attention to one considerable advantage of the small producer—high-quality crops.

The European Lesson

In the early days of my own marketing education, I had the opportunity to visit many small farms in Europe. I learned a great deal there. The farmers I visited were successful first of all because they worked hard. Second, they succeeded because they worked intelligently. They took advantage of their strengths and minimized their weaknesses. For example, they diversified in order to spread their production, income, and family labor over a longer season. They chose agricultural enterprises that fit together spatially or temporally; that is, a particular crop grown in a particular space would fertilize that plot of ground for the following crops. The periods

A selection of cabbages.

Garlic sold on the stem like a flower.

Make sure the customers know where to find you.

of heaviest work did not come all at once. These small farmers succeeded because they had access to modern production technologies scaled to their operations. But most important of all, they were economically successful because they understood how to compete in the marketplace with the most valuable product they had to offer—quality.

Standardization in food does not exist in Europe on the scale it does in the United States. Regional and varietal differences are treasured. Quality in food is demanded as much as is quality in other products. In Europe small growers are encouraged to produce a more carefully nurtured, high-quality product. What does that quality consist of? Many factors: Truly fresh produce. The tastiest and most tender varieties. The most careful soil and cultural practices needed to grow the crop correctly. The widest selection. The longest season. The personal touch. All are criteria in which a small farmer producing for a local market has a great advantage over bulk-grown, trucked-in foodstuffs. European farmers' success lies in concentrating on the areas in which the small farm excels.

I think the back-to-the-land movement of the 1960s, which was responsible for increasing the interest in organic growing, missed one important

A late fall farmers market in a large glass greenhouse.

facet of the old rural society—the passion for quality. The old peasant crafters found their joy in doing the job as well as they could do it. And then doing it better the next time. Whatever their field of endeavor, the goal was to create of an object of beauty.

Quality from Within

I have noted over the years that many people feel uncomfortable about hustling a product, about putting on the hard sell, and so forth. Those people can calm their fears right now. If their product is good, necessary, and really first-class, then they never need worry about finding customers. The negative image many folks carry of salespeople stems from the association of that profession with unwanted, unneeded, shoddy, poorly made, frivolous products that require fast talking and other unsavory skills to sell. Good growers should understand that they are not in the same category at all. Theirs is a first-class product: It is produced locally without polluting the environment; it saves energy because the food doesn't have to be transported across the country; the production of it stimulates employment and strengthens the local economy. A quality product always benefits the buyer as well as the seller. And that is how transactions between human beings should be, a mutually beneficial two-way exchange.

Ideas need to be sold, too, especially those ideas that are fundamental to good farming. Therefore, this chapter deals with more than just selling a product. It also deals with caring. Quality is the result of the skill of the producer coupled with care. Experience will provide the skills, but caring must come from within. Many customers may not be aware of real quality. The public is not always well informed about agricultural practices and their effect on the quality of food. It's the growers who must care about their agricultural practices and the repercussions, even though there may be no one forcing them to do so. Giving a damn and doing what is right are rewards in themselves. In the long run, producing a poor or deceptive product, or a less-good-than-you-can-do product, is harder on the producer than on the customer. The customer may only encounter it once or twice. The producer has to live with it all the time.

Real Carrots

Even though the customer may be adept at distinguishing quality in manufactured goods, this perceptiveness does not always extend to food crops because he or she often has no standard for comparison. To most people, a carrot is a carrot is a carrot. Well, to tell the truth, it isn't. And the difference is not just looks. Carrots can take up large quantities of pesticide residues when they are grown in soils that contain them. Carrots grown in soil with a low pH take up more lead, cadmium, and aluminum. Unbalanced soluble fertilizers modify plant composition. The lack of minor elements, limited by poor soil conditions, inhibits protein synthesis. There is

Communicating with customers.

definitely a biological value in food plants that is hampered or lost by inadequate soil fertility and growing conditions.

For the most part, scientific evidence on the subject of food's biological value is contradictory and incomplete.[1] It remains one of those areas where people are intuitively conscious of differences once they experience them, but tests for those differences are inadequate. There are documented differences in food quality as affected by growing conditions, but not enough to constitute "absolute proof"—as if absolute proof is ever possible with any biological concept. Perhaps we aren't yet wise enough to know what to look for. But the sensible shopper doesn't always wait for science. The idea is to lead science rather than follow it. I remember watching meticulous French customers shopping for vegetables in a rural village. Their standards, which had been honed by generations of awareness, were very high. If "a carrot is a carrot is a carrot," you could not have proved it by them.

Other ideas are spreading as well. Agricultural chemical technologies have been under heavy criticism since the 1960s. People are demanding safer food and more control over additives and residues. The idea is dawning that food must be judged not by its cosmetics but by its composition. The public is aroused because of a sense that the system they have been taught to rely on has betrayed them and will probably continue to betray them.

That system, which is composed of the regulatory agencies of government—such as those for environmental protection, food and agriculture, and product purity—has compiled a dismal and embarrassing record of failure with regard to safeguarding the public. Stories to that effect appear in the media

every day. Pesticides or additives sworn to be absolutely benign one day are banned as hazardous the next. Scientists upon whose knowledge we depend to make those decisions disagree vehemently among themselves over the safety of agricultural practices vis-à-vis the health of those consuming the food, the pollution of our water supplies, and the long-term productive capacity of the soil. But the average customer is no fool. The disturbing evidence can no longer be ignored. Something is amiss, and whether there is "absolute" scientific proof or not, a groundswell has begun. The emperor has no clothes, and the public, by doubting the adequacy of the regulatory agencies, has begun to make independent decisions to seek safer food.

We may seem to have wandered off the subject of marketing strategy, but not in practical terms. Quality is the linchpin of a small-scale grower's business. Once customers realize that certain producers care, that they are sincere, and that their word and their produce are dependable, they will patronize them faithfully. I ran my first market garden in Maine from 1968 to 1978 in a very unlikely location. The farm was 6 miles from a numbered highway (and the last 3 miles were dirt road). Marketing problems? I had none. Our produce set the quality standard, and we always had more demand than we could meet. Once a reputation for "real" food is established, there is no better advertising or marketing program. The market, as they say, will take care of itself.

The local organic grower has an additional marketing advantage that should be exploited. As the organic food industry has expanded, large-scale marketers, whose only interest is profit, have moved in. Despite the existence of a national certification program, deceptive practices have increased, and suspicions about whether the food is truly organic have become commonplace. The public will soon realize (and I for one will certainly encourage them to do so) that the best way to be assured of food quality is, in the words of the old quotation, "to know the first name of the grower."

CHAPTER EIGHT

Planning and Observation

When I began farming full-time on my own land in Maine, I was extremely fortunate to have as friends and neighbors Scott and Helen Nearing. The Nearings taught me a wide range of economic survival skills, but the most important were planning and observation. The Nearings demonstrated those two valuable skills at their best.

They were careful planners and organizers of the work to be done and the crops to be grown and always sought out the most efficient way to accomplish any task at hand. They were without a doubt the most practically organized country people I have ever met. In fact, I remember marveling that Scott was the one nonagenarian I knew with plans for the

With my neighbor Scott Nearing in 1974.

future farm project he would be working on ten years hence. Many of Helen and Scott's ideas and experiences as small farmers are described in their book *Living the Good Life* (Harborside, ME: Social Science Institute, 1954; reprinted, New York: Schocken, 1970).

Planning on Paper

I soon learned to plan ahead much more efficiently than I ever had—to set out the whole year's work on paper during the winter months and thus have a good grasp well in advance of what resources I might need, where they could come from, how I would acquire them, and how much time I might allot to each task. I organized a notebook into sections for each vegetable crop, for every year in the different rotations I was trying out, for fertilization records on each field, and so forth. There is no way to match the value of organizing and planning beforehand.

The Nearings were masters of observation. They meticulously recorded all the bits and pieces of data gleaned out of day-to-day farm activities—from which variety of lettuce wintered over best to what combination of ingredients made the most effective compost for peas. Some of their observations came from intentional comparative trials, but the majority came by chance—from keeping their eyes open and training themselves to notice subtle differences, where less perceptive observers would pass by unaware. In short, they never stopped learning and were wise enough to record what they noticed so it would be of use to them in the future.

Taking a cue from the Nearings, the first step, therefore, is to plan out your operation in detail. Let's go through this process step by step to figure out which crops to grow, in what quantities, and how to set it all up.

What to Grow

Depending on the market and the climate in your area, the possibility exists to grow anywhere from 1

to 70 or so reasonably common vegetable crops. Imaginative growers are rediscovering old crops every day. John Evelyn in his 1699 essay "Acetaria: A Discourse of Sallets" listed 77 vegetable crops, and those were just salad ingredients. The 48 vegetables I consider the most promising are listed in table 8.1 and are divided into two categories, major and minor.

TABLE 8.1. Most Promising Vegetables

Major	Minor
Asparagus	Arugula
Bean	Celeriac
Beet	Chinese cabbage
Broccoli	Collards
Brussels sprouts	Dandelion
Cabbage	Eggplant
Carrot	Endive
Cauliflower	Escarole
Celery	Fennel
Chard	Kohlrabi
Corn	Leek
Cucumber	Mâche
Garlic	Okra
Kale	Radicchio
Lettuce	Salsify
Melon	Scorzonera
Onion, bulb	Shallot
Onion, scallion	Turnip
Parsley	
Parsnip	
Peas	
Pepper	
Potato	
Pumpkin	
Radish	
Rutabaga (Swede turnip)	
Spinach	
Squash, summer	
Squash, winter	
Tomato	

One way to begin deciding which vegetables to grow is to write down in chart form any information that will help organize your planning. For example, I might begin by compiling a chart of the months when different vegetables could be available for sale if they were grown in my area. That chart should include the potential for extended availability of these crops if the growing season is supplemented by the protection of walk-in tunnels, the greater protection of a heated greenhouse, or out-of-season sales from a storage building.

Charts of the potential availability of crops for sale in my area, New England, might look like tables 8.2, 8.3, and 8.4.

Then, depending on whether I wanted to grow just seasonally or for an extended market, I would have an idea which crops could be available and when. The advantage of compiling this kind of information is that it stimulates thinking. It might suggest a specific course of action, such as a degree of specialization, perhaps. A wide variety of crops can be made available year-round. In many markets year-round production can help keep customers or acquire restaurant contracts. A look at the chart shows that many salad crops are capable of year-round production (see chapter 24).

The "A" crops in table 8.4 are the most potentially lucrative for the grower, but they are also the most expensive to produce. They need higher temperatures, requiring more heating costs and a more professional greenhouse, one that is taller and stronger for trellising.[1] They also are not actually year-round crops, although they are long-season. Only the most specialized producers plan on harvesting before April and after November.

The "B" crops can be grown in simpler tunnel greenhouses at lower temperatures. Some, such as mâche, parsley, scallions, spinach, and carrots, can be grown as fall crops with no supplementary heat at all. They can be harvested all at once before real cold sets in, or over a good part of the winter by providing just enough heat to keep them from freezing. (See chapter 24 for specific information on low-heat and no-heat winter production of these and other salad crops.) The decision depends on your market.

The most basic year-round greenhouse crop is lettuce. It is always in demand. Excellent varieties for winter production are available through the specialty seed catalogs. If you use an adapted variety, you can grow lettuce at low temperatures and plan winter harvesting on a regular schedule.

Production Size

This is a function of a number of other factors. How much land is available? How fertile is it? How many workers are involved? What kind of equipment is on hand? As I said earlier, I consider 1½ to 2 acres (6,070 to 8,095 square meters) of intensive production to be ideal. The decision about the size of a productive farm cannot be made in a vacuum. The relationship of size to all of the production and marketing factors must never be forgotten.

The market-garden layout will obviously be determined by the lay of the land, but in addition there are some general suggestions that are applicable almost everywhere.

Subdivision

No matter what size the field, it should be subdivided. One-hundred-foot-long (30 meter) sections are an efficient size for the scale of machinery to be used. A 5-acre (2 hectares) field, sectioned off, might look like this:

Ideally, the field will slope to the south. The beds run across the field. Each bed is 100 feet (30 meters) long. The spaces between the sections, which allow for access and turning a walking tractor/tiller at the end of each row, are 10 feet (3 meters) wide.

There are some solid reasons for subdividing. Ease of access, of calculating input and production information, and of general organization are just a few. The most important reason is management. Subdivision makes it easy to keep an eye on everything. Care is the

TABLE 8.2. Availability of Major Crops for Sale Fresh

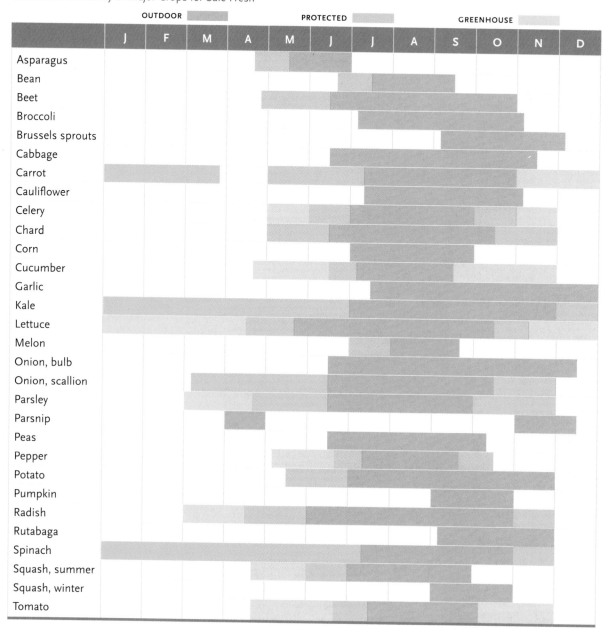

Layout and Crop Spacing

key, and nothing must be neglected. Subdivision helps to get you and your attention to every part of the operation. The crop that could easily be forgotten in the middle of a large field is more likely to receive care in a smaller space. No matter what the shape of the growing area, it should somehow be divided into workable sections.

The divisions above must now be progressively subdivided again. Just as a country is easier to comprehend when it is divided into states, counties, and towns, a garden is more comprehensible as sections and strips and rows. Each section is 100 feet (30 meters) by 30

TABLE 8.3. Availability of Minor Crops for Sale Fresh

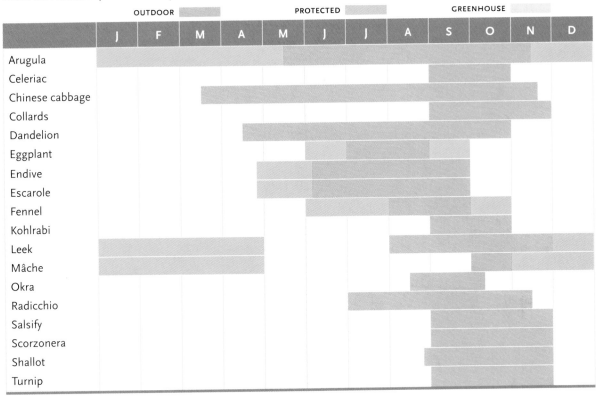

TABLE 8.4. Specializing in Fresh Salad Crops

feet (10 meters), or 1/14 of an acre (285 square meters). A bed is a part of a section 100 feet long by 30 inches (75 centimeters) wide. That creates 8 (30-inch) beds side by side in each section.

Foot traffic should be confined to the access paths between the beds in order to avoid soil compaction in the growing area. Thinking in terms of beds helps to make the production system more flexible. Any bed can be planted after harvest to a succession crop or to a green manure independently of the rest of the section.[2]

One useful hint: When you're tilling across a sloped field using a walking tractor, start at the uphill edge of the area to be tilled. Then, as you till the second and subsequent passes, it will be the upper wheels that sink slightly in the softened soil, creating a leveling and slightly terraced effect to the field. If you start at the bottom edge, this effect is reversed, and the tilt of the field is increased by the sinking wheels. It becomes harder to keep the tractor in line and to keep the field surface even.

For an operation specializing in salads I recommend a walking tractor with a 30-inch tiller and the wheels set on 42-inch (105 centimeter) centers. You can then till the 30-inch-wide growing area in one pass. The 12-inch (30 centimeter) access paths are cultivated with the 12-inch knife on the wheel hoe.

I began using 30-inch-wide beds separated by a 12-inch path years ago for certain greenhouse crops, and I soon adapted it to the field. I find it especially suitable when specializing in multicrop, fresh salad production where I am doing a lot of multiple harvests of low-growing crops and I need to be able to quickly move across the field. It is easy to step across a 30-inch-wide growing area, and you can straddle it comfortably if you wish while harvesting or transplanting.

Getting Good Seed

So many factors are important in vegetable growing that there may be disagreement with the values I ascribe to some of them. But I doubt if anyone will dispute the importance of good seed. Without high-quality seed, all the other activities are moot.

The grower will first be concerned with specific named varieties (or cultivars, as they are also called). At the start I suggest sticking with tried-and-true, locally adapted varieties from a dependable regional supplier. After the first year or two, the grower should have enough experience about what works and what needs improvement to begin selecting from numerous seed catalogs. With catalogs, it is mostly a case of learning to read between the lines. New seed varieties, no matter how highly praised, are always a risk. This is especially true with the larger commercial catalogs, which often select new cultivars for their ability to perform under conventional fertilizer/pesticide regimes. Organic growers will find, as I have, that the stable, older varieties often give more dependable results.

Some seed varieties, like some people, thrive best under specific conditions. That is not to say that you should avoid trying newly developed or hybrid varieties. Just never abandon a dependable old variety without being sure about the quality of its replacement. I have always found it rewarding to read specialty and foreign seed catalogs and then conduct trial plantings of promising varieties and even new crops. I have discovered a small but important number of my favorites this way. The number I trial is small because the seed companies are also testing and trialing varieties from many sources with far greater resources than I possess. For the most part their results are thorough and dependable. There always exists, however, the pleasure of a new discovery made on my own, and I heartily recommend the practice of seeking and trying.

Below are some criteria that you may want to consider when selecting varieties:

Eating quality. This is most important by far and includes the flavor, tenderness, and aroma of the vegetable, both raw and cooked.

Appearance. Color, size, and shape are also important but are secondary to eating quality.

Pest and disease resistance. This is useful where a problem exists; otherwise, choose a variety for its flavor and tenderness.

Days to maturity. This is obviously an important factor in planning early and succession crops.

Storage. Suitability for long or short periods in storage.

Vigor. This includes quick germination and quick growth.

Performance. Does the variety have vigor under a wide variety of conditions?

Standability. This describes non-cracking tomatoes, non-splitting cabbages, and so forth.

Ease of harvest. Carrots with strong tops are easier to pull, and beans held above the foliage are easier to pick.

Time of harvest. Various cultivars can extend your growing season.

Frost resistance and hardiness. These are spring and fall concerns.

Day length. There are short-day varieties for winter greenhouse production, and so forth.

Ease of cleaning. Some leafy greens hold their leaves high to avoid soil splash.

Convenience. This includes self-blanching cauliflower, non-staking (determinate) tomatoes, and other convenient growers.

Ease of preparation. This means long as opposed to round beets, round as opposed to flat onions, and so forth.

Adaptability. Many varieties winter over and provide early-spring growth.

Nutrition. Some varieties have higher levels of nutrients.

Marketability. This includes specialty, ethnic, and gourmet varieties.

Quantity

Quantity is the next concern. How much seed of each variety should be purchased? Planting techniques will affect this decision. If a majority of crops are transplanted as this book recommends, the grower will be able to get by with far fewer seeds than would be needed if plants were to be sown directly. Information on quantities of seed needed for direct-sown crops is given in most seed catalogs. At the start you might want to purchase extra seed just to be sure. A new seeder or a new setting could easily plant twice the seeds calculated until it is calibrated correctly. Nothing is as discouraging as running out of seed on a perfect spring planting day. The cost of seed for field crops is a small expense in most cases, and buying a little extra is good insurance for the grower.

If there are specific varieties or crops that become important to the farm's production, it is a good practice to purchase an insurance packet of those seeds from a second supplier. This is especially important with succession crops. Along with the first planting of the standard seeds, plant seeds from the insurance packet. If all goes well the extra seedlings won't be needed, but if the standard seeds don't perform well you will be covered and know where to order new stock. Be sure to set up credit accounts with favored seed firms so you can order seeds by phone quickly if there are any problems during the year.

It is wise to be covered in the same way when you are planning to use last year's seeds. Most of the time and for most varieties, year-old seeds that were stored properly (in cool, dry, and dark conditions) will work just fine. However, the savings are a false economy if a crop or a succession planting is lost because of seed failure. Be sure to obtain each year's seeds as soon as possible. Never wait until the last minute. Early planting dates have a habit of sneaking up on you; before you realize it, spring is here. Whether you purchase seed from a mail-order catalog or from a local supplier will depend on personal preference. What is necessary in either case is dependability: consistent quality and up-to-date information. If the seed stock for a certain variety is poor one year, it is important that the supplier inform its customers of this fact. Smaller growers are often not privy to this information, so it always pays to ask.

I suggest a further precaution. After a few years' experience, a grower should experiment with saving seed from open-pollinated varieties. *Open-pollinated* means that properly grown seeds will grow into plants that are true to type (unlike seeds saved from F1 hybrids, which typically are either sterile "mules" or revert to one of the hybrid's parent strains). For most crops the vigor and viability of seed grown under the careful cultural practices of this production system will far exceed those of purchased seeds.

I make this recommendation for another reason, too. I hope that the direction of present-day seed breeding, selection, and manipulation is favorable to the producer of high-quality vegetables. However, many older varieties are being abandoned or unnecessarily tinkered with. I now save seed from any open-pollinated varieties that I treasure for their eating qualities or excellent growth under my pro-

duction methods. Fortunately, organic seeds have become much more available, and organic seed breeding is commonplace. Seeds are the spark of the farm operation, and the more control the grower can exert, the more dependable the system will be.

When to Plant

The date of harvest depends on the date of planting. The span of time between the two may be longer or shorter depending on the effects of day length, weather, the aspect of the land, the crops grown, and many other growth-related factors. Although control is possible with protected cropping (tunnels and greenhouses), the earliest and latest unprotected outdoor crops are still important. They cost less to produce and they include many crops that are not usually grown with protection.

EARLY AND LATE PLANTING

The best information on your earliest and latest local planting dates will come from other growers, and not necessarily just the professionals. Good home gardeners are surprisingly astute about planting dates and other matters. More than once I have seen the experienced home gardener beat the pros to the earliest harvest. In truth, this is such a complex subject—and one that can be influenced in so many ways—that even the best growers are not doing all they could. Without a doubt the early outdoor production potential of many farms can be improved by paying attention to windbreaks, exposure, soil color, and other microclimate modifications.

As a further refinement there are specific cultural factors to take into account. Sweet corn, for example, is an important crop for most market stands, and earliness is what brings in the customers. But the planting date of corn can be pushed back only so far because of the limits of temperature. Corn won't germinate reliably or at all if the soil temperature is below 55°F (13°C). Yet corn seedlings will grow at or below that temperature. That is the factor usually overlooked. Pre-germinating corn seed or transplanting corn seedlings may be worth considering. Remember, we are not talking about the whole corn crop, rather just a few days' worth to catch the earliest market. Corn can be transplanted quite successfully if it is grown in soil blocks (see chapter 17).

SUCCESSION AND GREENHOUSE PLANTING

If a grower wishes to harvest a crop such as lettuce progressively throughout the year, it may seem logical to plant every week to keep up the continuity of supply. The logic is only partially valid. There may indeed be 52 planting dates during the year, but they will not be at seven-day intervals. The season of the year affects plant growth because of light, temperature, day length, and so on. Planting dates must be adjusted accordingly. Although these dates will vary with the geographic region and lay of the land, certain general patterns can be used as a guide.

The maturity time for lettuce is doubled and tripled for plantings from September through February. The spacing of the planting dates must reflect that reality. In order to harvest lettuce every week from early November through April, Dutch research has determined that the following planting schedule is necessary.[3] (Growers in the Southern Hemisphere can transpose all these dates by six months and obtain a reasonably close approximation of planting times.)

September 1–10	Sow every 3½ days
September 10–18	Sow every 2 days
September 18–October 10	Sow every 3½ days
October 10–November 15	Sow every 7 days
November 15–December 15	Sow every 10 days

Other trials have shown that seed-to-harvest times can be speeded up if lettuce transplants are grown under artificial lighting for the first three weeks.

Outdoor production has similar variables. In notes from my own earliest experience on my Maine farm, lettuce sown in an unheated greenhouse on March 1 and transplanted outside April 21 was ready for sale on about May 25, whereas lettuce sown April 1 in the same greenhouse and transplanted outside on May 1 matured on June 2. Remember, specific dates are only guidelines. I wish to stress the understanding of the concept and the general pattern. All growers need to compile information for the climate and conditions of their individual farms. In the long run trial and error will be the best teachers of specific planting dates. This is another area where keeping careful records can be so valuable to the success of your farm.

Despite all the best planning, climate is never consistent. Unusual extremes of heat and cold can make life difficult. One way to offset unpredictable weather is to grow more than one variety of any crop. Choose varieties for their slightly different performance under similar growing conditions, which allows you to "blanket" the ideal maturity date. The dependability of the harvest is more ensured by adding a comfortable flexibility that can absorb some of the shocks of climatic anomalies.

A Final Word

Another lesson I learned from the Nearings is the folly of working seven days a week. There is a strong temptation when starting out in farming, without the benefit of parents and grandparents having done much of the preliminary work years before, to try to do it all right now. Working non-stop day after day is not the best way to achieve that goal. You soon get stale and lose the sense of joy and pleasure that made farming seem so desirable in the first place. Scott and Helen taught me the importance of pursuing something different, at least one day out of the seven, no matter how much work needs to be done on the farm. Even in the midst of the spring rush, it always turns out that one day of change allows much more to be accomplished on the other days. Rest and reflection not only heal the body but help provide insight into how to get more accomplished with less work in the future so the same bind won't exist another year.

CHAPTER NINE

Crop Rotation

Most dependable agricultural practices are ages old. Crop rotation is a good example. Descriptions of the benefits of rotating crops can be found in the earliest Roman agricultural writings. The Greeks and, before them, the Chinese were also well acquainted with the principles of crop rotation. From his experience as a researcher at Rutgers, Firmin Bear stated that well-thought-out crop rotation is worth 75 percent of everything else that might be done, including fertilization, tillage, and pest control. In fact, I think this is a conservative estimate. Rarely are the principles of crop rotation applied as thoroughly as they might be in order to garner all of their potential benefits. To my mind, crop rotation is the single most important practice in a multiple-cropping program.[1]

The more variety in your crop rotation, the better.

In a word, crop rotation means variety, and variety gives stability to biological systems. By definition, crop rotation is the practice of changing the crop each year on the same piece of ground. Ideally, these different crops are not related botanically. Ideally, two successive crops do not make the same demands on the soil for nutrients, nor do they share diseases or insect pests. Legumes will be alternated with non-legumes. A longer rotation before the same crop is grown again is better than a shorter rotation. And ideally, as many factors as possible will be taken into account in setting up the sequence.

Space and Time

The key to visualizing crop rotations is to understand that two things are going on at once. Rotations are both spatial (crops move) and temporal (time moves). With both crop sequence and time to consider, there may be some initial confusion when considering complicated rotations. Hang in there.

A graphic representation of an eight-year crop rotation would look like figure 9.1. There are eight sections, with a different crop growing in each section. Now let's say we want to rotate these eight crops so that A follows B, B follows C, and so on. Adding arrows to the picture indicates the direction of rotation. In each case the letters represent where the crop grows this year. The picture for the next year would follow the arrows one space over and have A growing where B grows this year, with H growing in A's old place. The following year would have A growing where C grows this year, and so on.

When planning rotations I use 3-by-5-inch index cards with the crop names written on them. Some of you may wish to use a computer to display your planning options. I move the cards around as I try to determine the ideal sequence for the number of crops involved. Let's see how that works. With two crops, there is no problem. Take corn and beans. Corn is growing this year where beans will be next

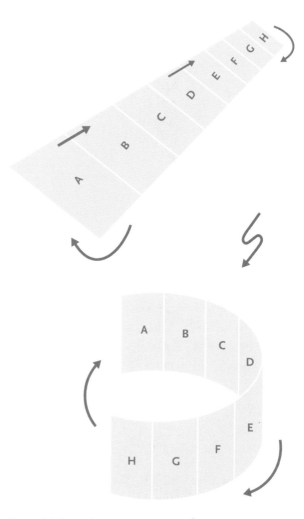

Figure 9.1. An eight-year crop rotation.

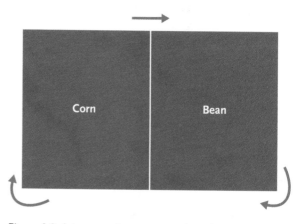

Figure 9.2. A two-year, two-crop rotation plan.

52

year and beans are growing this year where corn will be next year.

A three-year rotation expands that concept. Once there are more than two crops, a new factor is involved: order. There are now two possible sequences. The index cards can be placed: corn/beans/squash or corn/squash/beans.

Notice that we are not concerned with the number of ways in which three items can be ordered (of which there are six), but rather with the possible sequences. If we go to a four-year rotation, there are six possible sequences. An eight-year rotation offers 5,040 sequences.

Why Bother?

Decisions, decisions. Why bother with crop rotation? Because there are so many benefits to the grower from setting up a rotational sequence that exploits every possible advantage. Corn, beans, squash, and other crops all take different nutrients out of the soil. All respond to diverse fertilization patterns. All are amenable to specific cultivation practices. All may affect or be affected by the preceding or succeeding crop. Whenever the crop or cultural practices of the current year can be chosen to benefit a future crop, there is reason for bothering. In fact, whenever there is a choice between two or more ways of doing a job, one of them is usually the best way. Determined growers will take the time to think things through to optimize every aspect of their production.

Time spent planning a rotation is never wasted. Not only will you learn a great deal about important biological balances on the farm, but the results will be so effective in halting problems before they occur that you may sometimes have to remind yourself that a lot is happening. Very often farmers fail to take full advantage of a well-planned rotation, because rotations don't have any computable costs and because they work so well at preventing problems that farmers are not aware of all the benefits. Those benefits are, in a sense, invisible.

Insect, Disease, and Weed Control

Rotations improve insect and disease control by managing the system to benefit the crop. Monoculture encourages many pest problems, because the pest organisms specific to a crop can multiply out of all proportion when that crop is grown in the same place year after year. Pests are most easily kept in balance when the soil grows different crops over a number of years. A good rotation spaces susceptible crops at intervals sufficient to hinder the buildup of their specific pest organisms.

Rotations affect weed control in a similar way. The characteristics of a crop and the cultivation methods used to grow it may inadvertently allow certain weeds to find a favorable niche. A smart crop rotation will incorporate a successor crop that eradicates those weeds. Furthermore, some crops can work as "cleaning crops" because of the style of cultivation used on them. Potatoes and winter squash fit into this category because of the hilling practiced on the former and the long period of cultivation that is possible prior to vining for the latter.

Plant Nutrition

Rotations can make nutrients more available in a biological farming system. Some plants are more effective than others in using the less soluble forms of plant nutrients. The residues of these nutrient-extracting plants will make the minerals more available to later, less effective plants in the next sequence of the rotation.

In general, plants of a lower order of evolution have been shown to be better feeders on less soluble nutrient sources than those of a higher order of development. Lowly plants—evolutionarily speaking—such as alfalfa, clovers, and cabbages are more aggressive at extracting nutrients than more highly developed plants such as lettuce or cucumbers. Lettuce and cucumbers, I've found, don't feed well on less soluble mineral nutrients. Thus, in my rotations, the choicest spot and the finest compost is always saved for the lettuce and cucumber crops, and their exceptional quality has always repaid that care.

Manure

Rotations encourage the best use of organic soil amendments. Some crops (squash, corn, peas, and beans, for example) grow best when manure or compost is applied every year. Others (cabbages, tomatoes, root crops, and potatoes) seem to grow better on ground that was manured the previous year. Greens are in the former category, with the caveat that the compost should be well decomposed. Obviously, a rotation that alternates manured crops with non-manured crops will allow a grower to take these preferences into account.

Soil Structure

Rotations preserve and improve the soil structure. Different crops send roots to various depths, are cultivated with different techniques, and respond to either deeper or shallower soil preparation. By changing crops each year, the grower can make use of the full depth of the soil and slowly deepen the topsoil in the process.

Deeper-rooting plants of both cash crops and green manures extract nutrients from layers of the soil not used by the shallow rooters. In doing so they open up the soil depths, leaving paths for the roots of other, less vigorous crops. Deep rooters also incorporate mineral nutrients from the lower strata into their structure, and eventually, when the residues of these plants decompose in the soil, those nutrients become available to the shallow rooters.

Yields

Rotations improve yields not only in the many ways discussed above but also in subtler ways. Some crops are helped and some hindered by the preceding crop.

The University of Rhode Island conducted over 50 years of studies on the influence of the preceding crop on the yields of the following crop.[2] The possible reasons for this are numerous, and even after extensive study there is no general agreement on exactly what processes are involved. Some causes for the beneficial influence of preceding crops on subsequent crops are:

- Increase in soil nitrogen
- Improvement in the physical condition of the soil
- Increased bacterial activity
- Increased release of carbon dioxide
- Excretion of beneficial substances
- Control of weeds, insects, and disease

The injurious effects of preceding crops, which I aim to avoid by careful rotation planning, are produced by:

- Depletion of soil nutrients
- Excretion of toxic substances
- Increase in soil acidity
- Production of injurious substances resulting from the decomposition of plant residue
- Unfavorable physical condition of the soil due to a shallow-rooting crop
- Lack of proper soil aeration
- Removal of moisture
- Diseases passed to subsequent crops
- Influences of crops on the soil flora and fauna

Patterns

Despite a lack of agreement among researchers, certain patterns emerge from the studies I have read on good and bad rotational effects, as well as from my own observations:

- Legumes are generally beneficial preceding crops.
- The onions, lettuces, and squashes are generally beneficial preceding crops.

- Potato yields best after corn.
- For potatoes, some preceding crops (peas, oats, and barley) increase the incidence of scab, whereas others (soybeans) decrease it significantly.
- Corn and beans are not greatly influenced in any detrimental way by the preceding crop.
- Liming and manuring ameliorate, but do not totally overcome, the negative effects of a preceding crop.
- Members of the chicory family (endive, radicchio, and so on) are beneficial to following crops.
- Onions often are not helped when they follow a leguminous green manure.
- Carrots, beets, and cabbages are generally detrimental to subsequent crops.

These are merely patterns, not absolutes. Still, it is necessary to start somewhere. These patterns have been discerned through research on the influence of preceding crops on subsequent crops and from my own and other farmers' experience. Since these patterns may be soil- or climate-specific, they are offered mainly to indicate the kinds of influences to which alert growers should attune their senses. Whether universal or applicable only to a specific farm, these bits of wisdom can be valuable to the farmer who learns to apply them.

One Percenters

Whereas the rotation guidelines presented earlier in this chapter qualify under the category of standard crop rotation "rules," the patterns above belong more in the category of "suggestions, hints, and refinements." The effect of any of them on improved yield, growth, and vigor may only be 1 percent, an amount that may not seem worth considering to some. What must be understood is that a biological system can be constantly adjusted by a lot of small improvements. I call them one percenters. The importance of these one percenters is that they are cumulative. If the grower pays attention to enough of them, the result

will be substantial overall improvement. And best of all, these one percenters are free. They are no-cost gains that arise from careful, intuitive management.

One percenters may not always provide measurable results, but they have a definite influence. I have learned to pay attention and try to make use of them. Sir George Stapledon, an English grassland specialist and one of the wisest farm observers ever, was always aware of how much he did not know and how much science always misses or is ignorant about. He did not want that to limit his ability to act. I think that attitude is wise. Rather than not acting because we can't be certain, I suggest we try instead to apply what we hope we know. We should try to take as many intelligent actions as possible to incrementally improve our crops and then be attentive to what happens. Given our limited knowledge about all the interrelated causes and effects operating in the biological world, this seems to be the most productive attitude.

I continue studying ecological succession to find patterns for devising ever-better crop-rotation sequences. I am curious about the mechanisms of natural succession in disturbed ecosystems. I want to know the observed explanations for what follows what after fires or landslides or clear-cuts and so forth. Is this just a case of the availability of sun and shade, or are there progressive changes in the soil that dictate succession, and if so, how do they proceed? Are the pioneer crops merely opportunistic, do they just add organic matter, or are there other biological, chemical, or structural modifications that improve the soil's suitability to the needs of another crop? In other words, are there patterns, and can I replicate them? Because if there are identifiable patterns, I can employ similarities in the effect of vegetable or green-manure crops on the soil to create crop rotations that mimic natural laws. I suspect many of the observations made by growers over the centuries have picked up on a lot of this. But I also suspect there are endless incremental improvements to be made through further study.

A Sample Rotation

Before deciding what crops the As, Bs, and Cs of those earlier illustrations stand for, we must first collect a good deal of information. Toward that end, let's set up a sample rotation for our 2-acre vegetable farm. The following factors need to be considered.

NUMBER OF SECTIONS

A crop rotation works best if the rotational sections are all the same size. That goal is not always easy to achieve on the large farm, where whole fields are involved, but it should be manageable with 2 acres (8,095 square meters) of vegetables. For this discussion let us assume that we will be using 2 acres of land divided into 10 sections of ⅕ acre (810 square meters) each.

NUMBER OF YEARS

Just as two dozen crops don't necessarily mean a 24-year rotation, you should realize that 10 sections don't have to mean a 10-year rotation. Each section can be divided into two, three, or more separate and shorter rotational cropping plans. Possibly a legume/grass pasture could be included for a number of years in rotation. Each grower makes these decisions to suit his or her own situation. For now, let's say that the 10 sections will be managed as a 10-year rotation.

NUMBER OF CROPS

In the example we are working with at the moment, 24 major crops will be grown. To begin to plan where to grow each crop in the rotational sequence, we need to divide the crops, first by botanical classification. Table 9.1 shows how this can be done.

Table 9.1 is divided by vegetable families based on one of the first principles of crop rotation: Don't grow the same crop or a closely related crop in the same spot in successive years.

These lists make a good start, but more information can allow us to refine our rotation. It might help to divide up the crops according to more general

gardening categories. Table 9.2 classifies our rotation crops by type.

Although table 9.2 mixes up the botanical divisions, it adds valuable new information. Since more than one crop will be growing in some sections, categorizing the crops by type can help us decide which have similar cultural requirements and which (such

TABLE 9.1. Rotation Crops Listed by Botanical Family

APIACEAE
Carrot
Celery
Parsley
Parsnip

ASTERACEAE
Lettuce

BRASSICACEAE
Broccoli
Brussels sprouts
Cabbage
Cauliflower
Kale
Radish
Rutabaga

CHENOPODIACEAE
Beet
Chard
Spinach

CUCURBITACEAE
Cucumber
Squash, summer
Squash, winter

FABACEAE
Bean
Pea

LILIACEAE
Onion

POACEAE
Corn

SOLANACEAE
Pepper
Potato
Tomato

TABLE 9.2. Rotation Crops Listed by Plant Type

BRASSICA CROPS
Broccoli
Brussels sprouts
Cabbage
Cauliflower

FRUIT CROPS
Pepper
Tomato

GRAIN CROPS
Corn

GREENS
Celery
Chard
Kale
Lettuce
Parsley
Spinach

LEGUMES
Bean
Pea

ROOT CROPS
Beet
Carrot
Onion
Parsnip
Potato
Radish
Rutabaga

VINE CROPS
Cucumber
Squash

TABLE 9.3. Rotation Crops Listed by Space Needs

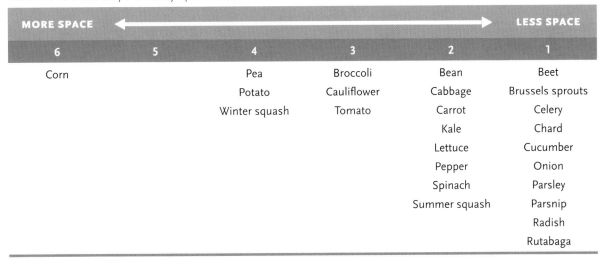

MORE SPACE					LESS SPACE
6	5	4	3	2	1
Corn		Pea	Broccoli	Bean	Beet
		Potato	Cauliflower	Cabbage	Brussels sprouts
		Winter squash	Tomato	Carrot	Celery
				Kale	Chard
				Lettuce	Cucumber
				Pepper	Onion
				Spinach	Parsley
				Summer squash	Parsnip
					Radish
					Rutabaga

as greens) might need to be harvested together for a specific market.

SPACE FOR EACH CROP

The fact that 24 crops will be grown in a 10-section rotation indicates that some of the crops do not need as much growing area to meet market demand as others. And that leads to one of the most interesting of the rotation-planning puzzles: how to meet the different needs of the market and still fit all of these disparate crops into a systematic crop rotation. This will be influenced by whether you plan to grow the widest possible variety of crops or plan to specialize in just the best-selling ones. The best place to begin is by deciding how much space or what percentage of the total cultivated area each crop needs in order to produce the right amount for market. We can determine those space requirements by creating six different categories, from the largest space needs to the smallest. From my experience, divisions for our 24 crops would look like table 9.3.

Now the index cards come into play. Each card will represent a section of the rotation. Write each of the names of the left-hand crops (those requiring the most space) on a separate card (or two cards, in the case of larger crops like corn). Take a pair of scissors and cut up proportional sections of other cards to represent the smaller areas needed by the right-hand crops. More than one of the smaller-space crops will occupy the same rotational section. Next, tape a number of them together in the space of one card. Whenever possible, put crops that are in the same family or require similar cultivation conditions together.

The Crop-Rotation Game

At this stage the arrangement and rearrangement of the cards is something like a board game. The rotation principles and patterns discussed earlier are the "rules." New rules are added as you become aware of them through experience, reading, and suggestions from other growers. Start the game by placing the cards on a flat surface and adjusting their positions to make up one rotational sequence or another. The aim is to determine if it is possible to grow all your desired crops on the land you have available and in the quantities you need, while at the same time satisfying all the rules. The winner is the sequence that comes closest to the ideal pattern, one that optimizes as many of the beneficial aspects of a crop rotation as can be achieved with your specific crops.

So let's give it a try. The two corn crops should not be side by side. Put one in the middle of the rotation and one at the end (figure 9.3), thus placing the corn crops as far from each other as they can be in a 10-year rotation. We know that potatoes yield best after corn, so put the potatoes in section 4. That move now naturally suggests a place for the tomatoes and peppers in order to create distance between them and the related potatoes (figure 9.4).

Since grain crops (corn) traditionally do well after legume crops (peas and beans), what if we precede the corn with the two legumes? (See figure 9.5.) Granted, there are the cabbage family crops sharing the bean section (and one of our patterns suggests that cabbages are negative preceding crops), but corn has been found to be the least affected by a preceding

detrimental crop. Further, the cornfield will likely be manured, helping to offset any negative effects.

Since beans are not affected too much by the preceding crop, let's put the often detrimental roots (carrots and beets) in front of them. Now a nuance can be considered. Onions have been shown to be a very beneficial crop before the cabbage family. We aren't growing enough onions to take advantage of that in whole sections, but they can still be effective depending on where the crops are placed in a section. In this case we can grow the carrots and beets where they will be followed by the beans, and grow the onions, as much as possible, where they will be followed by the cabbage family. What the heck? They have to grow somewhere, so it might as well be where they have a chance of doing some good. And since those cabbage family crops are in the bean section, the other brassicas

Figure 9.3. Corn has been placed in the crop-rotation plan.

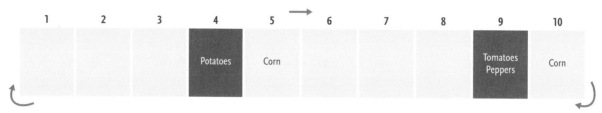

Figure 9.4. Potatoes, tomatoes, and peppers are now in place as well.

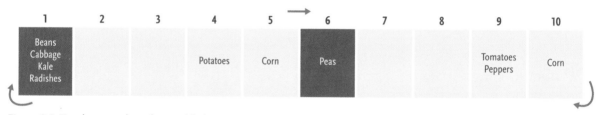

Figure 9.5. Two legumes have been added.

ought to be set apart from them. Section 7 would seem to be ideal (figure 9.6).

Now for the final two sections. Squash is a generally beneficial preceding crop, and it is well suited to growing with an undersown leguminous green manure (see chapter 10). Since that green manure would be excellent before the broccoli-cauliflower section, let's put the squash card in section 8. By default, the greens go to section 3 (figure 9.7).

Now let's see if this fits in with some of the rules we haven't considered yet. What about the crops that most benefit from manure or compost the same year

it is applied? The ideal situation has those crops alternating with the others. Not bad at all (figure 9.8).

I would suggest that the manure for the corn in section 5 could be omitted, since the pea crop will be finished early enough, even in my short season (120 frost-free days), to allow a leguminous green manure to be seeded and get well established by the end of the growing season. When tilled under the following spring, the green manure should provide more than adequate nourishment for the corn crop to follow.

If manure and compost are in short supply, some selective decisions will have to be made. It would be

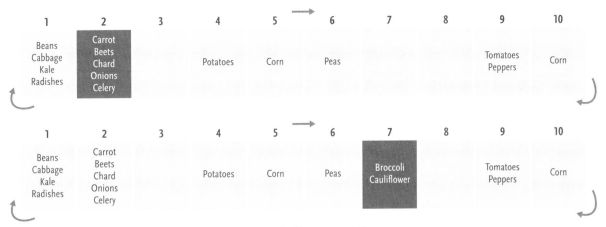

Figure 9.6. The addition of roots and onions (*above*), and of brassicas (*below*).

Figure 9.7. Squash and greens take their places.

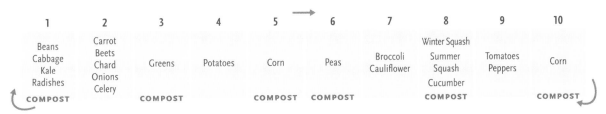

Figure 9.8. Adding compost/manure requirements to the rotation plan.

nice to aim for a manure application at least one year in three, but even that isn't vital. Instead, you have recourse to another management practice: the undersown green manures mentioned earlier. Those techniques will be discussed in the next chapter.

Multi-Year Crops

There is no problem including in the rotation those crops that need to remain in the ground for more than one year. They are assigned as many sections as the number of years they are to grow. Year 1 of a six-year rotation for four crops is depicted in figure 9.9.

Let's say crop A is strawberries. In year 1, only section 3 would be planted to strawberries. The other sections could grow green manure or any other crop unrelated to B, C, or D. In year 2, section 4 would be planted to strawberries.

The strawberries in section 3 are now in their second year and cropping. Section 2 can be treated again as in year 1. Then, in year 3, the rotation is off and running.

The strawberries in section 3 will give a second crop this year before being turned under. The section 4 berries are in their first year of cropping, and the Section 5 berries have just been planted.

The sequence continues in future years. If the As were a pasture or hay crop, the same system would prevail. One new section would be seeded each year and one old section would be tilled up and readied for the following crop. In fact this would be an excellent way to use a perennial deep-rooting legume like alfalfa or a legume-grass mixture (red clover / alsike clover / timothy, for example), to include some serious soil-fertility improvement in a rotation. The perennial soil-improving crops have been found to

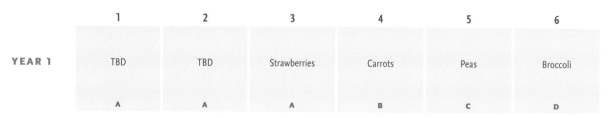

Figure 9.9. Year 1 of a six-year rotation for four crops.

Figure 9.10. In year 2, section 4 is planted to strawberries.

Figure 9.11. The third year of the rotation plan.

achieve optimal benefit to soil structure and produce increased organic matter if they can be left for three years. The virtues of alfalfa and many other green-manure crops are discussed in the next chapter.

Short Rotations

Not surprisingly, the reality is often not as perfect as the ideal. When there is no way to run a long rotation, you have to make the best of things. If a single crop dominates a large part of your production program, you may need to repeat it every other year—or, as in the case of some greenhouse lettuce production, even twice a year. In these short-rotation situations, changes should be introduced at every opportunity. That even includes changing the variety of the crop. Any slight genetic difference should be exploited if it adds diversity to the cropping program. Sowing a succession crop after the main crop can help. A green manure can follow or be undersown (see chapter 10). Mustard or rape—traditional cleansing crops for sick soil because they stimulate soil microorganisms—can be very effective so long as the dominant crop is not a fellow brassica.

In other words, aim for as great a variety of unrelated crops as possible in the span between related crops. Some growers advocate growing two consecutive crops (of, say, lettuce) followed by a longer break instead of alternating the crop with shorter breaks. I have not found that to be better, but I encourage a trial if the idea seems appealing. Other growers suggest that the more intensive the cropping, the more care must be taken to optimize all the growing conditions, especially by using extra soil-improving organic amendments like compost. I agree fully with that suggestion.

In some cases no rotation at all is recommended. Many old-time growers insist that tomatoes do best if planted every year in the same spot. They even recommend fertilizing them with compost made from the decayed remains of their predecessors. I once grew tomatoes that way for eight years in a greenhouse. In truth, they were excellent, and they got better every year. I do not grow field tomatoes that way now and cannot really defend my decision except to say that it is more convenient when they are part of the rotation. It could be that I am just uncomfortable about breaking the rules I have found to work so well with other crops. It could also be that I am unnecessarily limiting my options. I suggest that you try growing tomatoes (or any crop, for that matter) without rotation. Nothing is as stifling to success in agriculture as inflexible adherence to someone else's rules. With a little daring and imagination whole new vistas may open up. Remember, the aim of this farming system is independence, reliability, and sustainability. Any practices and attitudes that contribute to that goal should become part of the rule book.

A Tried-and-True Rotation

The 10-year rotation we just developed was meant as a teaching exercise. It may need refining for your operation. I'll now show you an eight-year rotation; it is a good one to conclude with, because I followed it for 10 years in the 1980s on a farm in Vermont. It has been well tested. I have thought about modifying it countless times but never have. Its virtues always seem to outweigh its defects, although that isn't to say it can't be improved. I'm sure it can. But it has been a dependable producer, and I offer it here as a tried-and-true example of a successful rotational sequence.

The goal of this particular rotation was to grow 32 vegetable crops in adequate quantities to feed for a year the community of 60-some people who ate daily in the dining hall of a school. Experience from both world war Victory Garden programs found it possible to grow a year's vegetables for 40 people on an acre. Thus the rotation that follows represents 1½ acres (6,070 square meters) of land. The salad crops not included here were grown in a separate small salad garden close to the kitchen.

1. **Potatoes** follow sweet corn in this rotation because research has shown corn to be one of the preceding crops that most benefit the yield of potatoes.
2. **Sweet corn** follows the cabbage family because, in contrast with many other crops, corn shows no yield decline when following a crop of brassicas. Second, the cabbage family can be undersown to a leguminous green manure, which when turned under the following spring provides the most ideal growing conditions for sweet corn.
3. **The cabbage family** follows peas because the pea crop is finished and the ground cleared by August 1, allowing a vigorous winter green-manure crop to be established.
4. **Peas** follow tomatoes because they need an early seedbed, and tomatoes can be undersown to a non-winter-hardy green-manure crop that provides soil protection over winter with no decomposition and regrowth problems in the spring.
5. **Tomatoes** follow beans in the rotation because this places them four years away from their close cousin, the potato.
6. **Beans** follow root crops because they are not known to be subject to the detrimental effects that certain root crops such as carrots and beets may exert in the following year.
7. **Root crops** follow squash (and potatoes) because those two are both good "cleaning" crops (they can be kept weed-free relatively easily); thus there are fewer weeds to contend with in the root crops, which are among the most difficult to keep cleanly cultivated. Also, squash has been shown to be a beneficial preceding crop for roots.
8. **Squash** is grown after potatoes in order to have the two "cleaning" crops back-to-back prior to the root crops, thus reducing weed problems in the root crops.

CHAPTER TEN

Green Manures

Not all crops are for sale. Green manures are grown not for cash but to contribute to the care and feeding of the soil. A green-manure crop incorporated into the soil improves fertility, but the eventual benefits are far greater than that.

Low-Cost Returns

Green-manure crops help protect against erosion, retain nutrients that might otherwise be leached from the soil, suppress the germination and growth of weeds, cycle nutrients from the lower to the upper layers of the soil, and—in the case of legumes—leave to the following crop a considerable quantity of nitrogen. Other contributions of a green manure are improved soil structure, additional organic matter, enhanced drought tolerance, and increased nutrient availability for plants.[1]

The value of green manures has been appreciated since the earliest days of agriculture. It should hardly

A vigorous green manure of rye and vetch on the uncovered plot of a movable greenhouse.

be necessary to extol their virtues here, yet the situation is similar to that of crop rotation. The full potential of green-manure use is still underappreciated and unexploited. Also like crop rotation, green manures represent a management benefit: They are farm-generated production aids that offer an excellent return from little effort or expense. Granted, the seeds for a green-manure crop may have to be purchased, but their inclusion in the crop rotation yields benefits far exceeding their small cost. When green manures are included in the overall soil-management program, the combination of green manures and crop rotation can result in a truly unbeatable vegetable-production system.

Growing green manures has traditionally been viewed as an either-or situation. You grew either a paying crop or a green manure. If the use of green manures means replacing a cash crop, then the lack of interest in them is understandable. There are other options. But first, let's review the general benefits of green manures.

Inexpensive Nitrogen

Leguminous green manures are a most economical and inexpensive source of nitrogen. The nitrogen is produced right where it is needed—in the soil. In fact, when leguminous green manures are used effectively and levels of organic matter are maintained, any additional application of nitrogen is often unnecessary. The symbiotic process by which leguminous plants fix nitrogen from the air depends on a number of factors for its success. First, the soil pH should ideally be between 6.5 and 6.8. Second, the proper rhyzobium bacteria for the specific legume must be present in the soil. A bacterial inoculant should be applied if there is any doubt. Inoculants for specific legumes come in both powdered and granular form and can be purchased from farm stores or seed catalogs along with seeds. Finally, a soil test for the trace elements molybdenum (Mo) and cobalt (Co), both known to be important catalysts for symbiotic nitrogen fixation, is often a worthwhile investment.

Humus

Every little bit of organic matter added to the soil helps add to the all-important store of humus. Humus, the end product of organic-matter decay in the soil, is the key to good soil structure, nutrient availability, moisture supply, and the biological vitality of the soil. Some forms of residues are more long lasting than others. Very young, sappy green growth will stimulate a lot of activity in the soil but will not contribute much, if anything, in the way of lasting humus. Old, dry residues take longer for the soil processes to digest but are more valuable in building humus reserves. A 2- or 3-inch (5 or 7.5 centimeter) growth of recently sown oats or clover is an example of the former. Brown, frosted, and dried-out cornstalks would be at the other end of the scale. A lush green manure is probably better mowed and left to wilt for a day before being incorporated into the soil, to help slow down what could be a too-rapid decomposition. The tough mature crops will decompose faster if they are chopped or shredded before they are incorporated in the soil. Most of the green-manure crops that we will be concerned with fall somewhere between these two extremes.[2]

Stable Nutrients

Plant nutrients can be lost from unprotected soil. During fall, winter, and early spring, when commercial crops are not in the field, not only will growing green-manure crops hold the soil against erosion, but their roots will capture and use available plant nutrients that might otherwise be leached away. Prevention of this waste is considered to be so important on most of the small European farms I have visited that the farmers think of the harvesting operation as having two inseparable parts: first, the harvesting of the crop, and second, the seeding of the land to a winter green manure. Seeding is done as soon as possible after harvesting.

In addition to the nitrogen nodules on the roots of legumes, green manures provide further contributions to the mineral nutrition of subsequent crops. The green-manure plants themselves, once decomposed by soil organisms, provide the most direct contribution. An indirect contribution results when the process of decomposition aids in making further nutrients available. Decaying organic matter can make available otherwise insoluble plant nutrients in the soil through the action of decomposition products such as carbon dioxide and acetic, butyric, lactic, and other organic acids. Carbon dioxide is the end product of energy used by soil microorganisms. Increasing the carbon dioxide content of the soil air as a result of the decomposition of plant residues increases the carbonic acid activity, thus speeding up the process of bringing soil minerals into solution.

Soil microorganisms are also stimulated by the readily available carbon contained in the fresh plant material, and their activity results in speeding up the production of ammonium and nitrate. Even soils naturally high in organic matter, such as peats or mucks, are improved by the incorporation of a green-manure crop, which makes them more biologically active.

Biological Subsoilers

The deep-rooting ability of many leguminous green-manure crops also makes them valuable as biological subsoilers. Where soil compaction exists, deep-rooting green manures can bring a startling improvement in subsequent crops, solely by penetrating and shattering the subsoil with their roots. This opens up the soil, permitting the crop roots to more easily reach lower soil levels, where they find greater supplies of water and nutrients. Studies have shown a considerable improvement in drought resistance and crop yields following lupines, sweet clover, alfalfa, and other taprooted green manures.

Overwintered Green Manures

There are three ways in which green-manure crops can be managed: as overwinter crops, main crops, and undersown crops. Green manures can be sown for overwintering after a market crop has been harvested. For example, in the crop rotation at the end of chapter 9, a leguminous green manure could be sown after the pea harvest, which would occupy the ground until it was tilled in the following spring prior to planting cabbages. The other option would be to plant a second market crop after the peas. In many cases that might be desirable, but the benefit from a wintered-over legume that provides ideal growing conditions for next year's crop is a strong incentive for growing it.

Main-Crop Green Manures

In this case the green manure occupies ground in place of a market crop during the growing season or, even better, for up to three years. If extra land is available, this is a highly recommended practice, and when the green-manure crop can be grazed by livestock, it serves a double purpose. If you prefer to put all your land into market crops, however, you must choose between the future benefits of the green-manure crop and the potential income from a market crop. Since this is a choice that usually goes against green manures, often at the expense of the soil, I recommend a third management option, one that allows you to have a leguminous green-manure crop and the cash, too: That option is known as undersowing.

Undersown Green Manures

Undersowing, also known as overseeding or companion seeding, is the practice of growing a green manure along with the market crop. When done correctly, undersowing provides the best of both worlds. It is established practice in small-grain growing. The clovers or other legumes are sown with or

My homemade multiple-row seeder.

Sweet corn undersown to forage soybeans.

shortly after the wheat or oats, for example, and grow slowly in the understory until the grain crop is harvested. In vegetable growing this practice was not common, to the best of my knowledge, but since the 1980s has begun to be seriously considered.

The advantage of undersowing is that the green-manure crop is already established at harvesttime. In my northern New England climate, winter rye is the only green manure that can be seeded after fall harvest. A legume cannot be established that late in the season. Since in my experience legumes are the most beneficial green manures, I try to use them whenever possible. The only way I can do that without taking land out of cash-crop production is to undersow them.

TIMING CROPS AND GREEN MANURES

The practice of undersowing is something like planting desirable weeds between the crop rows. In a way that is very similar to the relationship between weed competition and crop growth, the effect of the undersown plant—the deliberate weed—on the crop plant depends on the age of the crop. Weeds can overwhelm young crops if they both start at the same time. Weed research has shown that crops will do fine if they have an adequate head start. If most crops are kept weed-free for the first four to five weeks after establishment, later competition from low-growing weeds will have little effect on them. If we interpret that correctly, then the best crops for undersowing would be low growing, and the best sowing date for

Winter squash undersown to sweet clover.

green manures would be four to five weeks after the establishment of the crop plants, whether direct-sown or transplanted. My experience bears that out.

Where timing is important, there is a tendency to err on the safe side. Why not wait six weeks or more before undersowing the green manure just to be sure? The problem is that the balance is tipped too far in the other direction. Since the undersown "weed" is deliberate, I want to be sure it grows. If I wait too long before undersowing, the crop plants will be large enough to overwhelm the green manure. The trick is to undersow when the crop plants are well enough along not to be adversely affected by the undersowing but not so well established as to hinder growth of the undersown green manure.

How does this timing work out in actual practice? In my cool climate, where crops such as corn, beans, squash, and late brassicas are often not planted or set out until June 1, I find the Fourth of July to be just about perfect, year in and year out, as the date for undersowing those crops. Obviously, later crops or succession crops will have their own dates. In those cases the four- to five-week delay before planting the undersown crop may be shortened if the growth rate of the crop is more rapid due to the warmer soil. Once you acquire experience you can also judge timing by the size of the crop plants. The only crop for which I can give reliable size data is corn (maize). Undersowing once the corn reaches a height of 10 to 12 inches (25 to 30 centimeters) has proven most successful for me.

BEFORE UNDERSOWING

Successful undersowing requires a clean, weed-free seedbed. Sowing the green manure is no different from sowing the crop: When seeds are planted into a weedy mess they become the seeds for failure. I have often thought that another side benefit of undersowing is that it motivates the grower to pay attention to clean cultivation right from the start simply because there is one more reason to do so. Like any problem "nipped in the bud," weeds are easiest to control early in the season. The clean seedbed prepared for undersowing is a by-product of early weed control. At least three cultivations should be made prior to undersowing, the last one just a day or two beforehand.

The goal of the grower is to provide every opportunity for the undersowing to get well established without weed competition. Unless the garden has a lot of weed pressure, the canopy of the undersowing will join with the crop canopy to keep later weeds from germinating. The few that do pop through should be pulled before they go to seed. Occasional forays down the rows will keep these competitors from becoming a problem.

SEEDING THE UNDERSOWN CROP

I have seeded undersown crops both by broadcasting and by drilling with a multiple-row seeder. If the undersown crop is broadcast in the standing market crop, the seeds can be mixed shallowly into the soil with a flexible tine rake. If the undersown crop is drilled between the crop rows, the seeds are planted at the proper depth in contact with moist soil, where they are certain to germinate.

Thinking It Through

I experimented a great deal with undersowing when I was farming in Vermont, where I had a lot more land available. It is a fascinating technique for moving your market garden to a higher plane of sustainability once you are established and weeds are under control. Years ago I built a multi-row undersowing seeder by bolting five Earthway seeders together. (See the photograph on page 66.) Now that I am farming more intensively with rows closer together, I undersow with the same four-row or six-row seeders that work so well for sowing multi-row vegetable crops at close spacing, and I confine my undersowing to wider-spaced field crops like squash, pumpkins, melons, and cucumbers as well as between the rows of fall-harvested cabbage, cauliflower, broccoli, kale, and brussels sprouts, where the undersown cover of wild white clover prevents muddy conditions when harvesting in wet fall weather.

Sowing dates and equipment for undersown green manures should be as well thought out as those for the cash crops. Sowing dates should be marked on the calendar. The seeds should be ordered ahead. The equipment should be quick, simple to use, and in good working order. Figure 10.1, on page 71, is an illustrated flow chart that shows how the crop rotation in the previous chapter can be combined with undersown green manures. Obviously, green manures are most effectively employed when they are considered an important component of the crop-rotation planning.

There is another parallel between green manures and crop rotation that should be noted. Variety in green manures is as important as variety in the market crops. Because green-manure plants also have different faults and virtues that affect the soil and following crops in different ways, green manures should be rotated to include as many different varieties as possible. A great deal of research work has been done on multispecies "cocktail" mixes for cover crops usually containing 8 to 15 species from lists of grasses, legumes, and brassicas chosen to maximize diversity. The latest development on our farm, in order to create as much homegrown fertility as possible, is the sowing of half of our best vegetable land to a multispecies pasture mix every year and letting that be grazed over the summer by our laying hens. The following year that land is in vegetable crops, while the other half has been sown to pasture mix. (See chapter 14.)

In studying the undersown green-manure chart (figure 10.1), you will note that six of the eight rotational plots are undersown, a seventh is sown to legumes after early harvest, and only one—potatoes—is seeded to rye after fall harvest. The ground is never bare. The soil is always growing either a market crop or next year's fertility. For much of the summer, it is growing both!

Which Green Manures?

My choices of green-manure crops for different uses are:

- With tall crops—sweet clover, vetch, red clover, or alsike clover
- For sodlike cover—dwarf white clover
- For resistance to foot traffic while harvesting—dwarf white clover or vetch
- Before potatoes—soybeans or sweet clover
- Under corn—soybeans, sweet clover, or red clover
- Between rows of root crops—sweet clover or dwarf white clover
- Soil protection that will winter-kill—spring oats, spring barley, or, in warmer climates, a winter legume that will complete its growth in spring and can then be mowed off
- For the latest fall planting in cold climates—rye or winter wheat

In the milder European climate, mixtures of green-manure seeds are sown after harvest to provide late-fall grazing. In parts of Germany these mixes of species for a green manure are known as Landsberger Gemenge. A Landsberger mix commonly consists of two legumes, a grass, and a cabbage family crop. When a field of Landsberger is ready for fall grazing, it looks like a tossed salad for livestock. Sample mixtures might include:

- Oats, red clover, field peas, and mustard
- Wheat, white clover, purple vetch, and rape

- Rye, ladino clover, winter vetch, and oil radish

In order to become well established, they should be sown at least six weeks before the first fall frost.

Green-Manure Review

Green-manure varieties and combinations are endless and are not limited to the ones listed here. The varieties mentioned here worked for me as I developed the biological production technologies for my particular soil and climate. Instead of talking about specifics that are so often regional, I want to emphasize principles that are more nearly universal—not only because different parts of the globe require different green manures but because there are no hard-and-fast rules. Although it is possible to present the broad outline of a biological system inside a book, the fine-tuning that goes on within that outline is the province of the grower. The best innovations and improvements usually come from the grower and not from any chart or list, no matter how complete it supposedly is. Whatever an expert does or does not say should not limit your options. The more involved you become in taking charge and perfecting the system proposed here, the more independent, reliable, and sustainable your system will become.

Here are some considerations to keep in mind when choosing green-manure crops:

Time of seeding. Early, late, intercrop, undersown, overwinter, year-round?

Establishment. The ideal crop is easy to establish and grows rapidly.

Time of incorporation into the soil. How mature is the green manure? I often refer to full-maturity soil-improving crops as brown manures because they are higher in celluloses and lignins, which are resistant to quick decomposition and thus result in longer-lasting soil organic matter. What is the following crop—seed or transplant? Legumes turned under in the fall lose 70 percent of added nitrogen, but only 38 percent when turned under in spring. With a winter-killed green manure it may be possible to transplant the spring crop directly without incorporating the green-manure residues into the soil. The same can be done with overwintered green manures if you wait until they have matured sufficiently in the spring so they will no longer regrow after cutting.

Rotational fit. The green manure should not share susceptibility to diseases or insect damage in common with the crop plants.

Feed value. When a green manure serves as animal feed, manure is deposited on the soil, and fertility is enhanced even more.

Soil microorganisms. Rape, for example, stimulates the biological activity of the soil.[3] Soybeans improve scab control in potatoes.[4]

Beneficial insects. Some green manures can serve as nurse crops for useful insects. This is an emerging field of knowledge with much to be learned!

Cost. Is the seed expensive? Can it be easily produced on the farm? Will the crop yield both seed and feed? Will a less costly seed be as effective if it is managed properly?

Undersowing Legumes

Considerations when choosing an undersown legume include:

- Shade tolerance.
- Ability to grow with the crop.
- Effects, including competition, on this year's crop.
- Beneficial effects on next year's crop.
- Erosion control.
- Winter hardiness. In some situations a legume that winter-kills is preferable, to avoid having a vigorous residue in the way of an early-spring sowing.
- Weed control. Rapid growth and broad leaves are pluses.

Green Manures in Rotation

Undersown green manures can be used extensively within the eight-year crop rotation discussed at the end of the previous chapter. The following sequence has worked out very well in practice. Figure 10.1 shows how all the pieces can fit together.

Potatoes cannot be undersown easily if the cultivation method used is hilling. I have grown potatoes without hilling by planting at a depth of 6 inches (15 centimeters) and filling the furrow partly at first, then completely after the potato greens reach the surface. Vetch can then be planted as an undersown legume. If the green manure is to be established following the potato harvest, winter rye is probably the best choice as a green manure.

Sweet corn is undersown to soybeans because research shows that a soybean crop almost totally inhibits potato scab organisms in the soil. The soybeans also grow well in the understory of the corn and provide excellent weed suppression.

The cabbage family is undersown to sweet clover, which is one of the best leguminous green manures to turn under for next year's corn crop. It grows well under the cabbage family because it is a taprooted crop that does not seem to interfere with the more shallowly rooted brassicas.

Peas are not undersown but are followed by a mix of clovers as soon as the peas can be cleared. This combination of legumes grows until it is turned under the following spring, by which time enough nitrogen has been fixed to ensure a splendid crop of brassicas.

Tomatoes are undersown to oats or some other non-winter-hardy grass crop. Certain grasses have been found to be excellent preceding crops for legumes such as peas, since they produce an allopathic effect that suppresses grasses and other weeds, but not legumes. It is important to choose a non-winter-hardy cultivar so there will not be a mass of fresh green growth in the spring

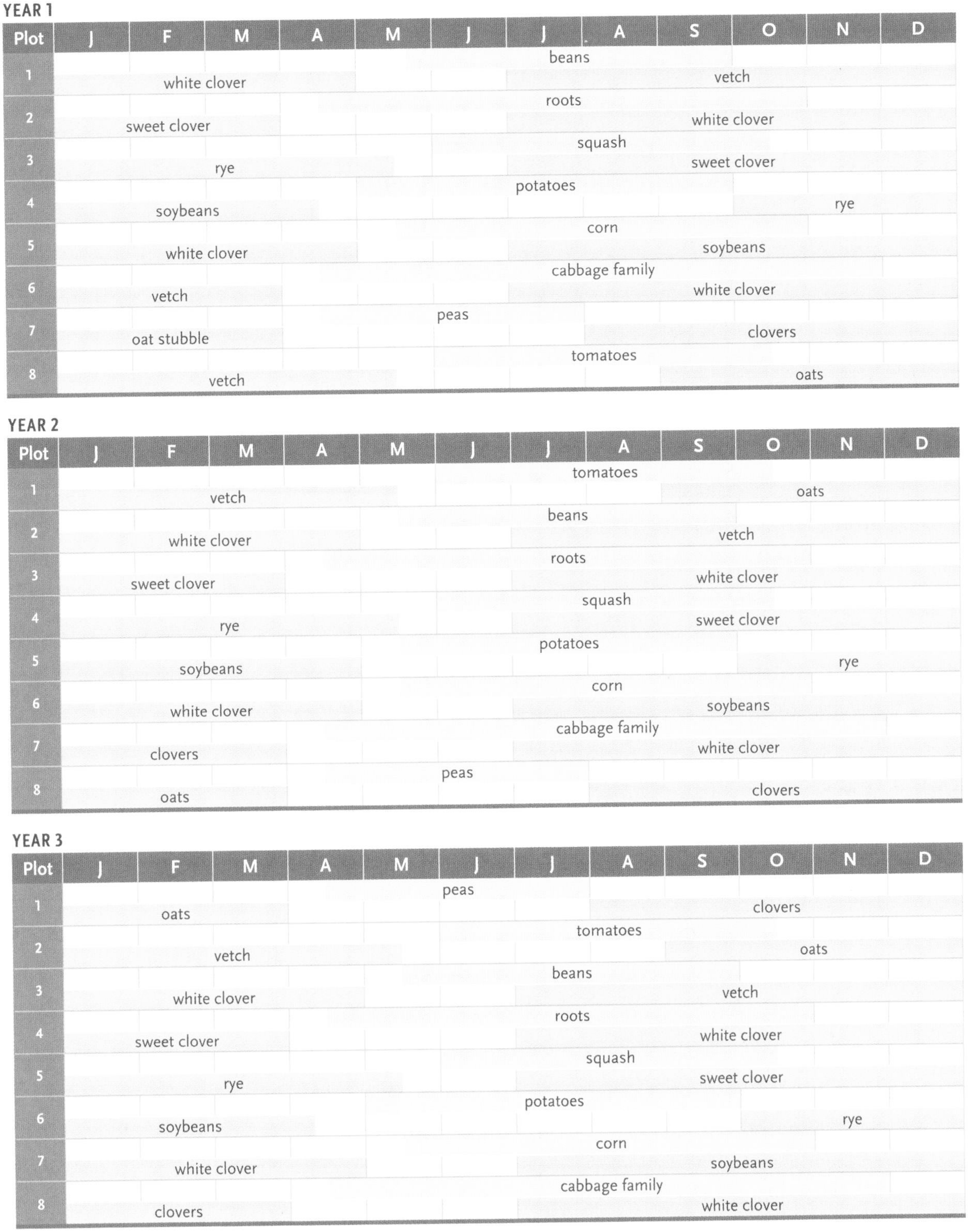

GREEN MANURES

Figure 10.1. A crop-rotation plan combined with a green-manure plan.

YEAR 1

Plot	J	F	M	A	M	J	J	A	S	O	N	D
1	white clover	white clover	white clover	white clover	white clover	white clover	beans	beans	vetch	vetch	vetch	vetch
2	sweet clover	sweet clover	sweet clover	sweet clover	sweet clover	sweet clover	roots	roots	white clover	white clover	white clover	white clover
3	rye	rye	rye	rye	rye	rye	squash	squash	sweet clover	sweet clover	sweet clover	sweet clover
4	soybeans	soybeans	soybeans	soybeans	soybeans	soybeans	potatoes	potatoes	rye	rye	rye	rye
5	white clover	white clover	white clover	white clover	white clover	white clover	corn	corn	soybeans	soybeans	soybeans	soybeans
6	vetch	vetch	vetch	vetch	vetch	vetch	cabbage family	cabbage family	white clover	white clover	white clover	white clover
7	oat stubble	oat stubble	oat stubble	oat stubble	peas	peas	clovers	clovers	clovers	clovers	clovers	clovers
8	vetch	vetch	vetch	tomatoes	tomatoes	tomatoes	tomatoes	tomatoes	tomatoes	oats	oats	oats

YEAR 2

Plot	J	F	M	A	M	J	J	A	S	O	N	D
1	vetch	vetch	vetch	tomatoes	tomatoes	tomatoes	tomatoes	tomatoes	tomatoes	oats	oats	oats
2	white clover	white clover	white clover	white clover	white clover	white clover	beans	beans	vetch	vetch	vetch	vetch
3	sweet clover	sweet clover	sweet clover	sweet clover	sweet clover	sweet clover	roots	roots	white clover	white clover	white clover	white clover
4	rye	rye	rye	rye	rye	rye	squash	squash	sweet clover	sweet clover	sweet clover	sweet clover
5	soybeans	soybeans	soybeans	soybeans	soybeans	soybeans	potatoes	potatoes	rye	rye	rye	rye
6	white clover	white clover	white clover	white clover	white clover	white clover	corn	corn	soybeans	soybeans	soybeans	soybeans
7	clovers	clovers	clovers	clovers	clovers	clovers	cabbage family	cabbage family	white clover	white clover	white clover	white clover
8	oats	oats	oats	oats	peas	peas				clovers	clovers	clovers

YEAR 3

Plot	J	F	M	A	M	J	J	A	S	O	N	D
1	oats	oats	oats	oats	peas	peas				clovers	clovers	clovers
2	vetch	vetch	vetch	tomatoes	tomatoes	tomatoes	tomatoes	tomatoes	tomatoes	oats	oats	oats
3	white clover	white clover	white clover	white clover	white clover	white clover	beans	beans	vetch	vetch	vetch	vetch
4	sweet clover	sweet clover	sweet clover	sweet clover	sweet clover	sweet clover	roots	roots	white clover	white clover	white clover	white clover
5	rye	rye	rye	rye	rye	rye	squash	squash	sweet clover	sweet clover	sweet clover	sweet clover
6	soybeans	soybeans	soybeans	soybeans	soybeans	soybeans	potatoes	potatoes	rye	rye	rye	rye
7	white clover	white clover	white clover	white clover	white clover	white clover	corn	corn	soybeans	soybeans	soybeans	soybeans
8	clovers	clovers	clovers	clovers	clovers	clovers	cabbage family	cabbage family	white clover	white clover	white clover	white clover

to impede early soil preparation and planting of the pea crop.

Beans are undersown to winter vetch. It is a dependable preceding green-manure crop for tomatoes.

Root crops are undersown to dwarf white clover (both in the paths and between the rows) because it will grow in the crop understory and because it provides good erosion protection for the soil over winter.

Squash is undersown to sweet clover in the empty strips between the squash rows. Beets, carrots, and other root crops grow very well following sweet clover. The onion crop, on the other hand, grows best with no preceding green manure, so plant onions in the strips that were occupied by the squash plants themselves.

The flow chart in figure 10.1 is an attempt to show visually the combination of crop rotation and undersown green manures in an eight-plot rotation over three years. The rotational sequence has the crops in plot 1 moving to plot 2, the plot 2 crops to plot 3, the plot 8 crops to plot 1, and so forth. If you look across the top half of the plot 1 strip, you can see over the three years how the beans are followed by tomatoes and the tomatoes by peas. If you look across the bottom half of the plot 1 strip you will see the undersown legumes that are tilled under prior to the seeding or transplanting of the crop and then undersown in the crop at the appropriate time. They remain through the winter until they are turned in the following year before the next crop.

Continuing to use plot 1 as an example, notice that, in addition to producing three crops (one crop per year over 36 months) here in my cool climate, that plot has also spent 26½ of those months covered with an undersown green manure. In milder climates two factors will change. The plot may be double-cropped (two crops in one season) or the list of potential green manure crops will be greatly expanded, as will the winter period during which the weather is mild enough for continued green-manure growth.

A discussion of the possibilities of basing the entire soil-fertility program on farm-generated inputs such as composts, crop rotations, and grazed pastures—no matter what your climate—is presented in chapter 13.

CHAPTER ELEVEN

Tillage

T illage is the general term for soil preparation in agriculture. It includes working the soil; incorporating lime, fertilizers, and manures; turning under green manures and crop residues; and any other mechanical processes involved in preparing the land for raising crops. The traditional implements are the plow, the disk, the harrow, and occasionally the subsoiler. Plowing with a moldboard plow is the method popularly associated with farming. In the process of plowing, the soil layers are turned over, either wholly or partially, depending on the adjustment of the plow. Since this operation alone does not produce a suitable planting surface or sufficient mixing action to fully incorporate fertilizers or organic materials, supplementary operations are necessary.

The depth wheels on the chisel plow can be removed for deeper penetration.

In order to come closer to the ideal of tillage, plowing is commonly followed by disking and harrowing. The idea is to loosen the soil; incorporate air, organic matter, and fertilizers; and remove weeds in order to prepare a clean seedbed. As a result of tillage, the air, moisture, temperature, chemical, and biological levels of the soil are modified. The intent is to optimize their effects on the growth and development of the crops.

Tillage operations are divided into those that work the soil deeply and those that work it shallowly. Deep tillage can go down as much as 18 inches (45 centimeters) using a subsoiler. Shallow tillage disturbs no more than the top 6 inches (15 centimeters) of the soil, and preferably only the top 2 inches (5 centimeters).

Deep Tillage

More and more scientific studies are pointing to subsurface soil compaction from plow pans and wheel traffic as a serious problem in crop production. The subsoiler is the original tool for deep-soil tillage, which ideally loosens the lower soil layers and breaks up hardpans and compaction layers without inverting the soil or mixing the subsoil with the topsoil. In addition, deep tillage can aerate the soil to a considerable depth—improving drainage, increasing rooting depth, making soil nutrients more accessible for the roots, and initiating a process of topsoil deepening, which greatly increases the fertility of the soil.

THE CHISEL PLOW

For our purposes the chisel plow is more effective than the subsoiler. The chisel plow dates from the 1930s, when it was conceived as a soil-conserving alternative to the moldboard plow. The chisel plow consists of a strong metal frame bearing a series of curved, soil-penetrating shanks (chisels) about 2 inches (5 centimeters) wide and 24 inches (60 centimeters) long that can be fitted with different tips.

When pulled through the ground, the chisels penetrate to depths of 12 to 16 inches (30 to 40 centimeters), but they do not turn over the layers the way a moldboard plow does. Rather, they simply lift and loosen the soil and break up hardpan and compacted soil.

This is not a small-scale implement. Using it once a year for deep tillage involves renting a tractor or hiring an operator. The latter is the simplest and most economical way to go if you live in an area where the services and equipment are available. If not, the solution might be for a group of growers to collectively purchase a small chisel plow and hire a tractor to pull it. If rocky land is involved, the tractor should have a front-end loader to help in collecting and removing rocks.

Many growers believe a chisel plow is suitable only for rock-free soil. In my experience that is not the case. I have used a chisel plow on both stone-free land in Texas and fairly rocky land in Maine, Massachusetts, and Vermont. It performed well in both cases. In fact, as a tool for preparing New England soil for vegetable growing, it is invaluable. The chisel plow finds rocks and brings them to the surface. You can then remove the rocks by rolling them into the tractor bucket on the next pass.

There is no need to despair if you find it isn't possible to get a chisel plow. There are other options. The hand tool (broadfork) suggested below and certain biological techniques will also do the job. The more attention you pay to improving pH, drainage, and organic matter while minimizing compaction, the less you'll need mechanical deep tillage. I do, however, recommend the chisel plow as an extremely valuable tool in the initial years of creating a fertile soil for vegetable growing.

After the first few years, though, you should be able to gain the same tillage effect with the roots of green manures. Deep-rooting green manure crops (alfalfa, sweet clover, lupines, soybeans, and red clover) are very effective at improving conditions in the subsoil. The deep rooting not only improves the soil physically by loosening it but also increases its fertility by bringing up more nutrients from the lower strata. The root channels remain long after the green manure has decomposed. They measurably help improve the soil's porosity and water-holding properties as well as preventing future hardpan formation.

THE BROADFORK

This two-handled deep-tillage tool is known by different names, but *broadfork* comes as close to describing it as any other. As with most agricultural tools, its genesis surely dates far back in agricultural history. It consists of a 24- to 30-inch-wide (60 to 75 centimeter) spading fork with a 5-foot-long (1.5 meter) handle at either side of the fork. The teeth on the fork are spaced 4 inches (10 centimeters) apart and are about 12 inches (30 centimeters) long.

Eliot using the broadfork in the field.

I first encountered this tool in the 1960s as the Grelinette, named after André Grelin, the French farmer who devised it. During the 1970s copies of the Grelinette began to appear in altered designs. As is often the case when a farmer builds a tool that is then redesigned by an engineer, Grelin's original design is far superior to the copies. Certain important nuances of Grelin's design are missing from the modern knockoffs, because they can be appreciated only through constant use and are not apparent on the drawing board. The copies, which are often fabricated entirely of metal to make them stronger, have a number of flaws.

First, the all-metal construction makes the tool too heavy. Granted, you can occasionally break a wooden handle; however, I much prefer working with a pleasant tool and being inconvenienced occasionally by a broken handle to working with a cumbersome tool all the time. A second mistake in the copies is the use of straight tines attached to the bottom of the crossbar. In the original Grelinette the tines are designed with a parabolic shape and curve down from an attachment point at the back of the crossbar. This difference is the key. The parabolic curve of Grelin's original design works with an easy, rolling motion. As the handles are pulled down, the tines curve under and lift the soil easily. With straight tines, a prying rather than a rolling motion is used, and you must muscle the soil upward using brute force.

USING THE BROADFORK

Hold the broadfork with the handles tilted slightly forward of vertical. Press it into the soil as far as possible by stepping on the crossbar, then pull the two handles back toward yourself in an easy rocking motion. Now lift the broadfork from the loosened soil, step backward 6 inches (15 centimeters), and repeat the maneuver. The tool is comfortable to use and makes the work pleasant.

How large an area can be managed with a broadfork? It is certainly scaled for use in commercial greenhouse vegetable production. I have used it outdoors on areas up to 1 acre (4,045 square meters) without feeling too much strain. The work can also be divided into sections and done only as needed prior to planting different crops. Anyway, there is no need to do every square foot (0.09 square meter). Just going down the row for widely spaced crops such as winter squash is sufficient. If the broadfork is used selectively during the planting season, even a 2-acre (8,095 square meters) garden is not unreasonably large for this tool. Some crops respond more than others. Sweet corn, root vegetables, and crops with extensive root systems such as tomatoes are greatly benefited.

The broadfork should be used prior to surface tillage, and preferably during the previous fall for sections of the rotation where the earliest crops will be planted. As with any tool it should be used with the eyes open: If there appears to be no difference in crop response, or a difference is apparent only on certain crops, then adjust when, how much, or how often you use the broadfork.

DEEP TILLAGE PROS AND CONS

The advantages of deep tillage are:

- Breaks up soil compaction[1]
- Provides soil aeration
- Aids the soil structure
- Improves drainage
- Extends crop-rooting depth
- Increases the range of soil nutrients available to plant roots
- Helps deepen the topsoil, which greatly increases soil fertility

The disadvantages of deep tillage are:

- Custom operators with chisel plow equipment may be hard to locate.
- A group of local growers may not exist to purchase a chisel plow collectively.
- In some soils the broadfork may be impractical on far less than 2 acres (8,095 square meters).

Solutions:

- You can use a rotary tiller as deeply as possible as long as you do not bring too much subsoil to the surface.
- When you're deep rotary tilling, be sure to mix extra organic matter with the soil to encourage an improved soil structure.
- Establishing the ideal biological soil conditions that favor bacteria and earthworms will improve soil structure and depth over time. Important practices include crop rotations, green manures, addition of organic matter, pH between 6 and 7, and adequate mineral nutrients.

Shallow Tillage

Shallow tillage is the preparation of the top few inches of soil. For many years I have used a rotary tiller for shallow tillage. It has many advantages over the traditional plow, disk, and harrow. First, it does the work of all three conventional implements in one operation. Second, it does the work at a speed that makes it considerably more efficient overall. And third, it does the job better.

THE ROTARY TILLER

In rotary tillage the soil is prepared by means of specially shaped soil-working blades (tines) that are rotated by a powered axle. A rotary tiller mixes and incorporates fertilizers, plant residues, and organic amendments (manures and composts) uniformly throughout the tillage depth and leaves them in contact with the greatest number of soil particles. This thorough mixing distributes organic materials and ensures the availability of minimally processed fertilizers.

The low-solubility mineral fertilizers I recommend in this book work best when they are mixed well into the soil rather than banded or layered. They then have the greatest contact with the soil acids and microbiological processes that make the soil nutrients available for use by plants.

The rotary tiller most easily incorporates manures, composts, and other soil amendments when they are spread on the surface of the soil before tilling. The relative density between the soil and the amendments affects how thoroughly the mixing can be accomplished. Mineral fertilizers, which tend to have the same density as soil particles, can be mixed in most uniformly. When organic materials are mixed in, the lighter stuff tends to remain higher in the soil profile, while the heavier material goes in more deeply. The rotary tiller's ability to mix both organic and mineral amendments uniformly throughout the soil profile is an important feature of this tool. It increases the fertility and biological activity of the soil that is so necessary to the establishment of a viable biological system.

There is evidence that the rotary tiller—or, for that matter, any soil-tillage equipment—can be detrimental if it is overused. The effect of rotary tilling on the soil is like that of using a bellows on a fire: It speeds up the combustion process. Extra aeration of the soil hastens the decomposition of organic matter, which can be good or bad depending on when it is done.

Spring tilling can have a beneficial effect. Early in the season the soil is cool and may well profit from tilling to help warm it up. Later in the season most of the soil will be undersown to green manures. When there is an early green manure, or when there are residues of a preceding crop to till under before a late crop, the extra air and thorough mixing will help decompose the residues faster and make the soil ready for planting of the following crop sooner.

ALTERNATIVES

The Europeans have used alternatives to rotary tillers for a number of years. My commercial vegetable-growing colleagues across Europe have universally replaced their rotary tillers with spaders—either rotary spaders or reciprocating spaders. These machines share the rotary tiller's ability to mix material throughout the soil profile while avoiding its often criticized tendency to beat up and overwork the soil.

The rotary spader looks something like a large-diameter rotary tiller but with spade-shaped blades on the end of long, curved arms. The spades move in a circumference three to four times greater than that of a rotary tiller. The spades also move more slowly through the soil than tiller tines and reasonably duplicate the soil-mixing action of a gardener using a spade.

Even more similar to hand-spading is the action of a reciprocating spader. The spades move up and down and then push backward by virtue of a camlike action that transfers the power of the engine to the spades. This action stirs and mixes the soil about as gently as can be done with a mechanical device. Both the rotary and reciprocating spaders move forward over the ground at a slower operating speed than rotary tillers. Consequently the area that can be covered per hour of work is not as great. Nevertheless, all the growers I have contacted who use spaders agree that the quality of soil preparation is sufficiently superior to more than make up for the increased time required.

The main difference between the rotary and the reciprocating spader is that the former is superior for turning under a heavy green manure directly, or when you wish to work the soil deeply. Some of the larger tractor-mounted rotary spader models advertise their ability to work the soil to a depth of 16 inches (40 centimeters). Whether such deep working is less detrimental with this tool or under certain cropping conditions, I cannot say. But my instincts lead me to be cautious. On a shallow soil, such as the one with which I started, any turning that went below 6 inches (15 centimeters) would likely bring up subsoil.

On the other hand, the reciprocating spader is acknowledged to be more gentle to the soil because its action more closely copies that of a gardener using a spade. It is the tool of choice in greenhouses and in cultures where heavy green manures are mowed before being incorporated into the soil. It might be at a disadvantage where large quantities of semi-decomposed crop residues are a major part of the soil-fertility program. The power sources for the greenhouse models are often modified so the engine runs on propane, since this fuel creates less air pollution in the greenhouse than gasoline or diesel.

Because of its method of action the reciprocating spader does not create a "plow pan," nor does it compact the soil underneath the wheels. It also deals with a stony soil, even the occasional large stone, much better than does the rotary spader. Most of the articles I have read from European sources

agree that the reciprocating spader is the all-around better choice. Growers like its beneficial effect on the soil: It increases root growth as well as improving water infiltration. The reciprocating spader also works organic matter into the top layers of the soil most effectively.

Another tool that incorporates organic matter even more gently and shallowly, by working horizontally rather than vertically, is the rotary power harrow. It is similar to having a number of small spike-tooth harrows that rotate on a vertical axis. The power harrow mixes the surface soil like your hand would if you put your fingers in a circular pattern, stuck the

Walking tractor tiller equipped with add-on depth-control roller. Photograph courtesy of Johnny's Selected Seeds.

tips of them shallowly into the soil, and turned your forearm. Although not suitable for primary cultivation (turning under sod or green manures), the rotary harrow incorporates any crumbly amendment like compost to a shallow depth very effectively. A power harrow is usually paired with a crumbier roller to maintain working depth and give a finished surface. Power harrows are now available for the walking tractor from a number of manufacturers.

A useful feature of the rotary harrow is the trailing crumbler-roller that allows the operator to precisely adjust the depth of soil working. There is now a new depth-roller option that can be added to the tiller on the BCS walking tractor. With that attachment the tiller now becomes an easily managed controlled-depth, bed surface preparation machine, and you can avoid the expense of purchasing a power harrow.

THE WALKING TRACTOR

A walking tractor is a two-wheeled power source. The rotary tiller in this system is powered by a 12-horsepower walking tractor. This equipment will give you enough power to do an excellent tilling job under almost all conditions. The walking tractor also has the flexibility to be equipped with a wide range of other implements such as seeders, rollers, mowers, hillers, pumps, and harvesters if the farm operation requires them. All of these attachments are available from the walking tractor manufacturer, but very often you can adapt the necessary implements from other sources. A walking tractor is my choice of power unit in my small-scale operation, for a few reasons: It is less expensive than a four-wheeled tractor; it is smaller and easier to work with when modifications or repairs are needed; and it is much easier to learn to operate. The walking tractor has long been the small farmer's best power source.

Of course, life might seem easier sitting on top of a powerful four-wheeled tractor fitted with large equipment. But the economics would not be the same. A well-built 12-horsepower walking tractor with a 26- to 32-inch- (65 to 80 centimeter) wide tiller

can be bought for a price that would purchase little more than the tiller for a large tractor. This same walking tractor/tiller also serves as the cultivator for the crops planted in wide rows.

Obviously, if the farm already owns a four-wheeled tractor and tiller, then by all means use it. The walking tractor is not better than a riding tractor, but it is perfectly adequate for the tasks you'll need it for. It is also more affordable, nicely scaled, and less expensive to maintain.

ADVANTAGES OF A WALKING TRACTOR/TILLER

Economics. The initial cost is less than that of a four-wheeled tractor, as are the operating costs.

Performance. Top-of-the-line models till as well as or better than many tractor-mounted tillers (except in old sod).

Flexibility. A walking tractor is basically a power source on wheels, and it is adaptable to many needs. It does the wide-row cultivation and hilling. Implements such as a water pump or a rotary mower can be run off the same unit, but I leave my machine set up just for tilling.

Simplicity. It is much easier to operate than a full-sized tractor, which means inexperienced helpers can quickly learn to use it, too.

Maintenance. It is less overwhelming and complicated than a full-sized tractor when repairs are needed. With its approachable scale, you will soon feel confident about making home repairs.

Lighter weight. It creates minimal soil compaction and leaves no deep wheel ruts.

Smaller size. It is far more maneuverable than a four-wheeled tractor, and less headland is required to turn it at the ends of the rows.

THE TILTHER

Even smaller and lighter than the walking tractor, the Tilther was designed as an electrically powered tiller for greenhouses. Many growers also find it handy for small areas in the field. The Tilther is 15

The Tilther mixing a soil amendment.

A homemade trigger for the Tilther.

inches (38 centimeters) wide (so it tills one half of a 30-inch/75-centimeter bed) and is driven by a side chain so it has no center gearbox to leave an untilled strip. It is powered by a cordless drill with a ½-inch (1.25 centimeter) chuck. I recommend an 18-volt or larger system. The Tilther has a stainless-steel cover over the tines with a flange extending to the rear, which limits the tilling depth to 2 inches (5 centimeters).

Tillage: The Future

Now that I've presented my best options for deep and shallow tillage, what about the possibility of reducing tillage, or eliminating it altogether? I think the idea is worth serious consideration. But it has to be done in a way that is more efficient rather than more complicated for the grower or it won't happen. I am not necessarily opposed to tillage. Where the results favor it, I continue to use it. Continuously tilling in organic matter over the years has allowed us to deepen our initial shallow topsoil to the present 10 inches (25 centimeters). But I am always on the lookout for any crop-specific benefits (yield, plant health) that might result from other techniques. There has long been a tradition among European organic growers to till no deeper than 2 inches (5 centimeters). In our greenhouses we now mix amendments (compost, alfalfa meal) into the top 2 inches of the soil with the Tilther. The results of this surface cultivation are very encouraging.

I am obviously interested in whether non-tillage can work in the field on the scale of a commercial market garden where, as in this system, extensive use is being made of green manures. I continue to experiment. While the general concept is workable, it's the specifics that need refining. One option we have tried for cleaning up the stumps, surface roots and leaf residues of cabbage, cauliflower, and broccoli is to mow and chop the field just below the surface. We use an old flail mower with "Y" knife blades, from which I have removed the depth roller, so the blades can penetrate one inch (2.5 centimeters) into the soil. (We call it the Earth Tickler.) A single pass leaves the field covered with a 1-inch- (2.5 centimeters) thick shredded mulch of brassicas and undersown clover mixed with soil. We can plant directly through that. Other options include using specially designed cultivators to clean the surface, growing winter-killed green manures that are easily brushed aside in the spring, or killing the previous growth with solarization or occultation (see chapter 19). The choice will depend on soil type, climate, and crops to be grown.

I am always impressed when I remove a straw mulch. The improved soil structure from the extra earthworm activity emphasizes how effective nature can be when I don't interfere.

A 1945 issue of *The Land*, a journal published by the Friends of the Land in the 1940s and '50s, reported on research at the South Carolina Agricultural Experiment Station involving no-till green manures.[2] Corn was planted into narrow furrows in clover sod. When the corn was 12 inches (30 centimeters) high, the clover was killed by cutting it just below the surface with special sweep cultivators. The clover was then left in place as a mulch.

Many related investigations have been done recently. USDA research in the northeastern United States uses overwintered hairy vetch as both a cover crop and a mulch. The vetch is mowed with a flail mower late enough in the spring so it won't regrow. The residues are left as a surface mulch for transplanted crops.[3] The eventual improvement in soil fertility from a surface-mulched green manure is the same as if it were tilled into the soil, and it offers the additional benefit of retaining moisture while it is being used as a mulch.

Two Australian researchers have devised a no-till green-manure fertility program they call Clever Clover.[4] Using separate fields for summer and winter production, they rotate both the vegetables (summer and winter crops) and the mulch varieties (subterranean clover and alfalfa). On one field the residues of subterranean clover, which grows during winter and dies out in the spring, serve as a summer mulch for transplanted crops. In the fall, depending on the variety, the clover either can be reseeded or will reseed itself to grow next year's mulch. The alfalfa, on another field, is mown every six weeks in summer. In autumn it becomes dormant, and winter vegetable crops are transplanted through the alfalfa residues. Come spring, the alfalfa begins growing again.

It is important to establish a vigorous cover crop so that weeds will be smothered. In all of these trials, plant diseases and pests were significantly reduced by the surface mulch technique.

Although I have been successful over the years in creating fertile soils with the mechanical tillage systems described in this chapter, I am not complacent.

A better technique is always out there waiting to be found. My preference for biology over technology makes me hopeful that new techniques will involve replacing mechanical solutions with biological solutions, in order to mimic the healthy, natural structure of undisturbed soil.

CHAPTER TWELVE

Soil Fertility

The main problem of permanent fertility is simple. It consists, in a word, in making sure that every essential element of plant food is continuously provided to meet the needs of maximum crops; and of course any elements which are not so provided by nature must be provided by man.

—CYRIL HOPKINS

I learned my first important lesson in soil fertility from the US Department of Agriculture, albeit by default on their part. It happened in 1966 when I began growing vegetables on rented land in New Hampshire. Since I had limited farming experience at that time, I eagerly read everything I could find on the subject. The USDA, in an article about fertilizer nutrients in agricultural production, stated unequivocally that nitrogen was nitrogen and phosphorus was phosphorus. They said that it did not make any difference to the plant where the nutrient came from. Nitrogen from manure and nitrogen from a bag of store-bought fertilizer were the same. To me, that was welcome news indeed.

Bring on the Free Manure

I was not aware at the time that the reason for these pronouncements was to discredit the organic farmers who claimed that manure or compost produced superior plants. In my naïveté, what I saw was a chance to save some money. I would not have to buy fertilizers! Just down the road was a horse farm with huge piles of rotted manure that the farmer was

giving away, even delivering free to anyone who wanted some. Chicken manure, which the article said was high in phosphorus, was available from another neighbor. I figured that if the USDA experts said there was no difference between the elements in manure and those in store-bought fertilizer, that was

Soil improvement through fresh seaweed.

good enough for me. I went with the manure and started a couple of acres of vegetables with what I assumed was the assurance of the USDA that everything would work out just fine.

And work out fine it did. During the three years I farmed that place, I had the best vegetables anywhere around. Not only that, but they got better every year. In fact, the old-timers were coming and asking the new kid how he did it rather than the other way around. Obviously, soil fertility was a function of a number of factors, and they did not have to be chemically processed or cost a lot of money to be successful.

When I did need to purchase nutrients, I continued to take the USDA at their word that elements were elements. I purchased unprocessed minerals such as rock phosphate. In the long run they were less expensive. Since they weren't water-soluble or subject to leaching, I could apply enough at one time to last a number of years. The lesson: Food for plants does not need to be prearranged in a factory. Nutrient availability is a result of biological and chemical soil processes that are stimulated by the agricultural practices I learned to use and trust—crop rotations, green manures, and animal manures. This biological system comes full circle. Each practice aids another, and the result is synergistic.

Building the Soil

To build a fertile soil, I focused right from the start on five amendments that I supplied as raw materials:

Organic matter. Compost or manure applied at the rate of 20 tons per acre (18,145 kilograms/4,000 square meter) every other year as a general rule.

Rock phosphate. A finely ground, natural rock powder applied every four years (quadrennially).[1] There are two forms—hard rock phosphate, containing 33

Crab shells are another fertilizer resource on the Maine coast.

percent P$_2$O$_5$ and colloidal phosphate containing 22 percent P$_2$O$_5$. I prefer the colloidal form, but other growers will make an equal case for the hard rock.

Greensand marl (glauconite). An ancient seabed deposit containing some potassium, but principally included as a broad-spectrum source of micronutrients. Applied quadrennially. Dried seaweed is another popular (although more expensive) source of potassium and micronutrients. It breaks down more rapidly and has the additional benefit of stimulating biological activity in many soils.

Limestone rock. A ground rock containing calcium and magnesium that is used to raise the soil pH. Sufficient lime should be applied to keep the pH within the range of 6.2 to 6.8.

Specific micronutrients. Elements such as zinc, copper, cobalt, boron, and molybdenum are needed in very small quantities but are absolutely essential for a fertile soil. They will usually be adequately supplied if the grower has paid attention to pH and organic matter. The need for supplemental application of micronutrients is best gauged through careful soil testing and grower observation. In many cases boron is the one element to need amending. Obviously, if a soil test indicates that one of the micronutrients is already well supplied, that supplement will not be needed.

Two Ways to Fertilize

Although this is a chapter on soil fertility, I am first going to discuss philosophy. That may be unconventional, but it is crucial to an understanding of the supplements recommended above. There are two basic philosophical approaches to fertilization:

Feed the plant directly. This involves using soluble fertilizers so the nutrients are "predigested" for plant use without the need for the natural soil processes.

Feed the soil and let soil processes provide for the plant. This involves creating and maintaining the optimal conditions of a fertile soil, under which a healthy soil-plant economy can exist.

In the first case, the farmer provides plant food in a "predigested" form because the soil processes are considered inadequate. A symptom—poor plant growth—is treated by using a temporary solution—soluble plant food. In the second case the farmer makes sure that the soil processes have the raw materials needed to be not just adequate but exceptional. The cause of poor plant growth—lack of sufficient plant food in the soil—is corrected by providing the soil with the raw materials needed to produce that plant food.

Natural Processes

Although we are dealing with agricultural techniques, we can't ignore the patterns of thinking that

Compost is the best fertilizer.

lead toward choosing one agricultural technology over another. These thought patterns stem from different points of view about the "natural system" that governs plant growth. Some questions:

- Are natural processes so inefficient that we can do better by taking over their roles, even though the energy cost of such a choice is high? Or can natural processes provide all that we need if we work to enhance them?
- Is it wise to rely on a crop-production system that is totally dependent on purchased materials involving great cost, supply networks, and safety considerations over which the farmer has no control? Or is it preferable to create a farm-generated system that relies on minimal quantities of off-farm products and maximum enhancement of the soil's inherent fertility?
- Is it acceptable to add only enough nutrients to get a crop? Or is it more worthwhile to try to provide all the known and unknown nutrients and growing conditions to allow the plants to grow at their optimum?

I have encountered many responses to these questions. There are always some growers who say, "Natural processes and growing optimum be damned. I just want to grow the crop with the least possible effort and deal with any problems later." Unfortunately, later problems, when they occur, are not limited to low yields, but involve insects, diseases, and poor crop quality. Other purchased products—pesticides, fungicides—are then used to deal with the new problems. Since agricultural systems are interconnected, one action leads to another, and one problem begets a subsequent problem.

My own position on these issues is that I simply do not know enough to tamper with the natural system, and I have no desire to do so. I am an admirer of the intricate cyclical systems of the natural world, and I prefer to study them in order to make less work for myself, not more. Even if I thought I knew every-

thing, I would rather let it be done for me by the real experts. The real experts in this case are all the processes that take place in a fertile soil—the interrelated activities of bacteria, fungi, dilute soil acids, chemical reactions, rhizosphere effects, and countless others we are unaware of.

My attitude toward the natural world is one of respect for a marvelously efficient system. If I attempt to feed the plant directly, I am in effect deciding that I can do a better job. On an infertile soil, where the system is working poorly, maybe I can. But on a fertile soil, the system can do a better job on its own. Therefore, my responsibility as a farmer is to add to the system the ingredients necessary to support a fertile soil. Those basic raw materials are organic matter and minerals in the form of powdered rock. So don't buy finished products. Buy the few raw materials that cannot be farm-produced, and let the soil processes finish the job. Not only does that policy make good sense agronomically, it is also the most successful, most practical, and most economical approach.

How It All Works

Let's say we start with an infertile soil. If we take the off-farm approach and add soluble fertilizers, a good crop can usually be grown. The soil serves merely as an anchor for plant roots, and the majority of the food for plant growth is provided by the fertilizer. The soil remains infertile, however, and the fertilizer application will have to be repeated for every crop. The situation is similar to helping a student by providing the answers to the test. The result may be a good grade, but the help will have to be given every time.

If, on the other hand, the second approach is chosen, we try to create a fertile soil by adding those ingredients that distinguish a fertile from an infertile soil. The fertile soil will then do what fertile soils do naturally—grow exceptional crops. To continue our student-and-teacher metaphor, this second process is like providing the student with the raw materials of knowledge (good books and study habits) so the student can develop the ability to excel on exams without help. I think most readers will agree that this second approach is the preferable choice in education. It is also the best choice for plant nutrition.

Feeding the Soil

The things that turn an infertile soil into a fertile soil are minerals and organic matter. If these are provided, the soil can excel on its own merits. Instead of a temporary crutch that must be provided time after time, a process is established that becomes self-sustaining. A fertile soil, like an educated mind, is a cumulative process, and with care it is capable of continuous improvement.

There are two sources of nutrients for the soil: the remains of previously living organisms that make up the organic matter, or humus, and the finely ground rock particles that constitute the mineral portion of the soil. Nutrients from both sources are made available through the biological and chemical actions that take place in the soil.

ORGANIC MATTER

For best quality and best growth, vegetables require the richest soils of all farm crops. And that richness has to be real. Not stimulants, but what British farmers so aptly call "a soil in good heart." Organic matter is the key to "heart" in a soil. The best book I ever read when I started out was not technically about farming. *Soil Microbiology*, Selman Waksman's classic text (see the bibliography), deals with the life in the soil. That book influenced my approach to soil fertility more than any other source. Waksman wrote from the point of view of one who had studied all the different life processes in the soil and how they affect nutrient availability and plant growth. His information opened my eyes to the marvelous world of living organisms under our feet and to the importance of organic matter for the well-being of that world. The quantity and quality of organic matter is the foundation for the microbiological life in the soil. This microbiological life grows and decays, solubilizes minerals, and liberates carbon dioxide as part of its life processes.

Organic matter also opens up heavy soils to make them more easily workable and binds a sandy soil so that it holds water better. In short, the organic-matter portion of the soil is more than simply a source of plant food and physical stability. It is also the power supply, so to speak. Organic matter is the engine that drives all the biological (and some of the chemical) processes in the soil.

Although raw organic materials such as crop residues can be added directly to the soil, it is often better to compost them in a heap. Sheet composting in the soil involves biological processes that preclude crop growth for a period averaging two to four weeks or more, depending on how resistant to decay the material is. In an intensive vegetable-growing system, that "soil time" would be better spent growing the next crop. For example, where early peas are to be followed by a succession planting, I will remove the pea vines to a compost heap rather than turning them under. The next crop can then be planted or transplanted immediately.

COMPOST

I have the highest regard for composted organic matter as a long-term soil builder. The crumbly, dark, sweet-smelling product from a heap of assorted plant residues mixed with straw is the finest compost of all. Well-made compost has been shown to have plant-growing benefits far in excess of its simple "nutrient analysis" and to be an active factor in suppressing plant diseases and increasing plant resistance to pests. Producing quality compost is the most important job on the organic farm. A lot of the problems I see on farms I visit could be solved by making better compost.

On the small scale I recommend making heaps inside enclosures of hay or straw bales (constructed by laying up bales like bricks, two or three bales high,

We make compost in hay-bale enclosures.

Finished compost under a protective cover.

Turning compost with a small PTO manure spreader.

as walls to contain the heap). I make the interior dimensions 12 feet (3.7 meters) wide and as long as I need to. This insulation of the bales keeps the compostables moist and warm right to the edge while still letting in adequate air. For more air, leave spaces between the ends of the bales. The bales usually hold together for two years (if baled with wire or plastic twine), and the hay or straw then becomes an ingredient in future heaps. I alternate 3-inch (7.5 centimeter) layers of dry brown materials with 1- to 6-inch (2.5 to 15 centimeter) layers of green ingredients—thinner for moist ingredients that might mat (such as young grass or clover mowings, or packing-shed wastes), thicker for loose, open materials (pea vines, tomato plants, and so on). I sprinkle a thin layer of topsoil over the green layer. If I'm adding montmorillonite clay, I sprinkle it on the straw layer. Since the ingredients are enclosed by the bale walls, they decompose right up to the edge. We turn the pile once after it is a year old by putting it through a manure spreader to

thoroughly mix all the ingredients. I protect the turned heaps with a compost cover to keep the sun from drying them out and the rain from leaching them. I use a covering material that is opaque to light but allows both air and moisture exchange. This system makes consistently high-quality compost.

Composted animal manure has long been one of the staples for soil improvement in vegetable growing. Manure can be either produced on the farm or purchased in truckloads from neighboring farms or stables. It may be given away or sold. Either way, you will usually pay for the trucking. There is an assumption that organic agriculture and use of manure as fertilizer are synonymous. That is not the case. It is organic matter (and not necessarily manure) that is so vital for soil improvement. Once it has decomposed in the soil or the compost heap, almost any addition of organic matter can be as effective as any other. That includes autumn leaves, straw, plant wastes, spoiled hay, or other locally available materials. Further, the green manures and cover-crop rotations discussed in chapter 10 are equally viable methods for adding organic matter to the soil in lieu of using off-farm supplies. Composted manure is a wonderful soil improver. By all means use it. But if manure is not available, you can plan your soil-improvement practices around many other sources of organic matter (chapter 13).

MINERALS

The mineral nutrients we are adding—limestone to keep the pH in the most favorable range for biological activity, colloidal phosphate (or rock phosphate) to supply phosphorus, and greensand for a broad range of micronutrients plus some potassium—are raw materials for the soil. (If greensand is unavailable, a dried seaweed product like kelp meal will fill the same bill.) Colloidal phosphate and greensand are considered to be unavailable sources in a "feed the plant" system because they are not highly water-soluble. But in practice their nutrients are made available for plants as a by-product of the soil's biological processes, which are stimulated by the cultural practices—high levels of organic matter, adequate moisture, soil aeration, green manures, crop rotations—mentioned so often in this production system.

To optimize the growth of crops, a grower must first optimize the workings of the biological "factory" in the soil. That is just what is done by all the cultural practices mentioned above. I can't stress enough the vital interrelations involved in this process; how much the benefits of these cultural practices are cumulative and synergistic; how much one makes another one work better. The cyclical concept is reaffirmed. No matter what the topic, we return again and again to cycles, because a sound biological system is a cyclical system.

Why Do It This Way?

But why do it this way? Why not just use water-soluble chemical fertilizers? Because plant quality is dependent on a balanced availability of nutrients.

The advantage of using the basic rock minerals in their natural form is that they are made available by the biological fertilizer factory of the soil at the rate plants need them. The two systems are harmonious. They evolved together. The correlation between availability of nutrients and their use by growing plants is a function of soil temperature, air temperature, moisture levels, and diurnal variation. Both plant and soil processes are affected simultaneously. During warmer, moister periods, plants grow better and nutrients are made available faster.

Thus, with natural rock minerals there are no problems caused by an excess supply of a soluble nutrient that upsets the mineral balance of the plant. Since there are no excesses, there is similarly no leaching of soluble nutrients such as nitrogen and potassium, which can be washed out of the soil to the detriment of both the groundwater and the farmer's pocketbook. Phosphorus doesn't leach out easily like nitrogen and potassium but rather becomes tied up in a soil that hasn't been programmed to release it. Either way, when water-soluble nutrients are used to "feed the plant," they do not contribute to a lasting soil fertility.

Other Rock Powders

Since the major structural component of soils is finely ground rock particles, from which bacterial action and plant roots extract mineral nutrients, some agriculturists have proposed amending the soil with rock powders other than lime, phosphate rock, or greensand. The suggestion makes sense.

Finely ground rock powders (usually waste products from quarrying operations)[2] add to the soil a material that approximates the composition of highly fertile, unweathered "young" soils. Soil scientists classify soils as young, early maturity, late maturity, and old. Young soils provide large amounts of essential plant nutrients because easily acquired minerals from fresh surfaces (the unweathered primary mineral particles) are abundantly available. In older soils the weathering has already either partially or totally taken place, and the nutrients are no longer being liberated in such abundance, if at all. According to experimental work that has been done on this subject, a number of factors determine the performance of rock powders as soil amendments.[3] These include the type of rock, the fineness of grind, the type of soil, and the type of plant.

TYPE OF ROCK

The first important variable is the type of rock. Over the years, trials have been conducted with volcanic dusts and pumices, granites, feldspars, and basalts, among others. Certain volcanic products have shown promise. Some granites have proven high in usable potassium, and biotite was the best of the feldspars. But the most researched and recommended have been the basalts.

The basalts are well-balanced rocks from the point of view of supplying soil nutrients. Basalts weather more easily than granites because they contain less silica and more calcium and magnesium. Soils derived from basalts are rich in clay and iron oxides and are usually very fertile. Basalt dusts are produced in large quantities as a result of trap-rock crushing operations, and at present they are a by-product looking for a use. European rock powder research has focused on basalt dusts in recent years, and they are used as an amendment in certain European biological farming systems. I suspect that in the future other rocks may also be appreciated as slow-release carriers of specific nutrients. An ideal product may someday be formulated that consists of a tailored blend of many different rock powders.

FINENESS OF GRIND

The second most important effect on the ability of rock powders to supply nutrients to plants is the size of the particles. The finer the grind, the greater the surface area of rock particles from which nutrients can be extracted. Other conditions being equal, the larger the surface area, the greater the availability of minerals. You can get an idea of the importance of particle size to mineral availability from the following

statistic: A pound (455 grams) of average rock in a solid cube would have a surface area of about 30 square inches (195 square centimeters). But when ground to a 300-mesh powder (very fine), the surface area is increased to some 16 million square inches.

FEEDING POWER

The activities of a plant's root system through contact with the mineral particles of the soil constitute its feeding power. This is another factor in determining the availability of nutrients from less soluble sources. Studies have shown that plants of the lower order of evolution, botanically speaking, have a better ability to extract less soluble minerals from rock sources than those plants that are more highly developed.[4] Examples of strong feeders are cotton, okra, apples, peaches, berries, roses, alfalfa, clovers, kale, cabbage, cauliflower, and radishes. Some weaker feeders are cucumbers, lettuce, sunflowers, grasses, and mints.

It would seem logical, then, to use rock powders to fertilize those crops and green manures that have been shown to utilize them most effectively, then turn the crop residues and green manures into the soil to make their nutrients available for subsequent crops. Or as one last interesting option, rock powders could be made on the spot as a by-product of getting rid of rocks. Powerful tractor-mounted machines are available that will crush rocks right in the soil. The machine could be set to crush progressively finer every few years when it was used again. It is appealing to think about turning the disadvantage of a stone-filled soil into the advantage of a long-term, slow-release source of minerals for plants.

Cool-Weather Cures

There are times of generally unfavorable growing conditions when the grower may want to use a temporary stimulant for plant growth. Let's take one common example. It is usually true that using soluble nitrogen or phosphorus to get plants growing in a cool spring may increase the bulk of the crop. Research has shown, however, that other nutrients (zinc, for example) are also immobilized by cool conditions and are unavailable to plants. Hence, the increase in bulk is an increase in quantity without quality, because the composition of the plant is imbalanced. Where a grower wants to stress food quality as a marketing tool, nutrient imbalance is unsatisfactory.

If growing conditions are inadequate, then that is where the improvement should be made. The answer to cool conditions is to cure the problem by providing climatic protection such as walk-in tunnels or low covers for early crops. In the warmer conditions created by this protection, the natural soil-nutrient mobilization processes will be able to function without artificial stimulation.

For field crops such as corn, where climatic protection is not practical, I recommend two valuable human attributes—patience and confidence. Patience because all will turn out well in the end, and confidence because it is often difficult to persevere when at first things look bad. Although the corn in a chemically fertilized field may be taller and greener early in the season, I guarantee that the corn from the biologically fertile soil will equal or surpass it by harvesttime.

It is a little bit like getting up in the morning. We can begin the day with some sort of stimulant or drug in order to "get going," but we pay the price later on through fatigue and continued reliance on the stimulant. Or we can accept the normal rate of mobilization of human energy that eventually results in a dependably productive day. There are a number of "natural" products on the market (liquid seaweed or fish emulsion are examples) that claim to offer plant stimulation without producing an imbalance. These products appeal to our human inclination to look for the magic bullet, a secret potion that will make everything work better. In my opinion these products have mostly a psychological benefit. They make the grower feel more secure because something has been done, whether it was necessary or not. Unquestionably, they can be helpful at times when things go wrong despite your best efforts. But I encourage you to use them as an occasional tool, not as a continual crutch. In most cases, you would be better off spending that money to build up long-term soil fertility.

Natural Reserves

In many cases the use of mineral amendments may not be necessary. If the soil has adequate trace elements, or if the manure is from animals fed a trace-mineral supplement, the micronutrient concern may be avoided. I do, however, recommend having some soil tests and tissue analyses made initially just to be sure. If the field has been heavily fertilized with superphosphate for many years past, there will usually be sufficient residual phosphorus for many years to come. Although the potassium content of the greensand is useful, it may not be necessary if manure is available. Further, the average agricultural soil has natural potassium reserves of from 20,000 to 40,000 pounds (9,070 to 18,145 kilograms) of potassium in the top 6 inches (15 centimeters) of each acre. Since I consider the usable soil depth for root feeding to be 24 inches (60 centimeters), and since the subsoil is equally rich in potassium reserves, the aggregate amounts are considerable.

Sandy soils do not contain as much native potassium, so they should be treated differently. But even these will often have adequate stores of potassium if they are considered to a depth of 24 inches (60 centimeters). Sandy soils may need heavier applications of organic amendments both initially and on a maintenance basis, not only to raise their potassium level, but also to improve their structure and water-holding ability. Ways to provide extra livestock manure for the farm at low cost, at no cost, or even at a profit are suggested in chapter 13. Obviously, any outside sources of organic material such as autumn leaves or manure from stables and racetracks should be investigated.

In some cases it might be a good idea to include an extra year or two of a small grain undersown to clover in the rotation, if only to have the straw for a humus-building soil amendment. Or three years of alfalfa might be included in the rotation and the four cuttings per year fed to livestock, composted, used as a mulch, or added directly to the soil and tilled in. On soils of lower initial fertility or on excessively sandy soils, one or more of these extra steps will be needed. The technique can vary as long as the goal is clear: long-term dependable fertility, not short-term plant stimulation.

Maintaining a Fertile Soil

A most important point to understand when following a feed-the-soil philosophy is that it is not necessary to apply every single year the amount of fertilizer required by the crop. Under a long-term approach, once a fertile soil has been established, it only needs to be maintained for plants to grow well. Obviously, in order to maintain and to improve the fertility of the soil, enough nutrients should be added to at least replace what is lost through erosion, leaching, and the sale of crops. If green manures are employed as assiduously as I recommend, though, there will be minimal losses from erosion and leaching, since the green manure roots will both hold the soil and use nutrients as they become available.

Once a truly fertile and productive soil has started to function, all that is technically necessary is to replace the equivalent of what leaves the farm in the produce that is sold. For example, if you wish to grow corn, you do not need to apply the entire amount of the nitrogen (N), phosphorus (P), and potassium (K) necessary for the corn crop each crop year. Adequate levels of these nutrients are available in the soil as a consequence of cultural practices.

How is that possible? Well, once a workable crop rotation has been established, not only is there a high level of fertility in the soil ready to assist the corn crop, but the preceding brassica crop has been undersown to a legume—white clover—which will be turned under in preparation for the corn. The crop residues from the brassica crop are also incorporated (they account for over 75 percent of the crop

mass that grew). If there is manure or compost available, that will more than complete the package, but it is not absolutely necessary. Using this system, I guarantee a first-class corn crop.

To look at it from another point of view, all that will be removed and sold from the corn crop are the ears. They represent less than 10 percent of the total nutrients in the corn plant. The rest of the plant is returned to the soil. If you calculate the amounts of P and K physically removed from the farm by the sale of corn, it would amount to approximately 17 pounds of P and 19 pounds (8 kilograms) of K, given an average 4-ton-per-acre (3,630 kg per 4,000 square meters) yield. The cabbage crop preceding the corn, if it gave an average yield of 20,000 pounds per acre (9,070 kilograms per 4,000 square meters), would have removed about 6 pounds (2.7 kilograms) of P and 50 pounds (23 kilograms) of K.

Taking those two heavy feeders as representative crops, let us assign 12 pounds (5.4 kilograms) of P and 35 pounds (16 kilograms) of K as the average removed from each acre of land each year. That amounts to 48 pounds (22 kilograms) of P and 140 pounds (63.5 kilograms) of K removed every four years. Our quadrennial mineral fertilization, which consists of ½ ton (455 kilograms) of colloidal phosphate containing 80 pounds (36 kilograms) of P and ½ ton of greensand containing 140 pounds (63.5 kilograms) of K (in addition to the micronutrients), makes up for those withdrawals. Supplement those figures with the 40 tons (36,290 kilograms) of manure or compost that may have been applied over the same four years—manure containing an additional 200 pounds (91 kilograms) of P and 400 pounds (181 kilograms) of K—and it is obvious that the soil is gaining both biological and mineral fertility.

The other major nutrient that hasn't been mentioned thus far, nitrogen, I do not consider to be in short supply. Nitrogen is available from the leguminous green manures, from crop residues, from non-symbiotic nitrogen fixation, and as a component of the manures or compost. These sources alone are sufficiently nitrogenous to cover all demands.

Biological and Mineral Activities of the Soil

The majority of plant nutrients are most available in the pH range of 6.2 to 6.8.

Soil microbial life is most active within that pH range.

The nitrogen fixation by legumes and bacteria is also most effective within that pH range.

Most soils contain immense reserves of potassium, which become available if the soil is biologically active.

Organic matter is the fuel that makes all soil processes work.

Many trace elements are crucial not only to the quality of the crops grown, but also to the optimal functioning of both the symbiotic and non-symbiotic nitrogen-fixation processes. However, it may not be necessary to supply them as inputs once a vigorous soil biology is established.

Finely ground rock powders (usually waste products of the rock-crushing industry) can be valuable soil amendments in a biologically active soil. European farmers who have explored these concepts consider basalt rock to be highly effective.

The more finely ground the supplemental rock minerals, the greater their surface area and the more effectively their nutrients can be liberated by bacterial action in the soil.

The most complex systems are the most stable. For example, the best compost is a broad mixture of ingredients. A compost of varied plants, weeds, garden wastes, and manure is better than one made only of barley straw or pine sawdust.

Fertile soil, healthy seedlings.

Self-Sustaining Soil Fertility

I arrived at these feed-the-soil concepts over many years of observing how best to grow plants and conduct the business of agriculture. But they are not new. Nor am I their only advocate. The ideal of sustainable soil fertility has been understood for ages and has been expressed by many writers. Perhaps the most forceful champion in this century was Cyril G. Hopkins, chief agronomist and eventually director of the Illinois Agricultural Experiment Station from 1911 to 1919. Hopkins had his own feed-the-soil philosophy, which he called the Illinois System of Permanent Fertility. He advanced his ideas both in his best-known book, *Soil Fertility and Permanent*

Soil Amendment Recommendations to Get Soil Fertility Started

Initial Application

(applied prior to year 1)

For a soil initially of low fertility:

20 tons/acre (18,145 kg/4,000 sq. m) manure or compost
1 ton/acre (910 kg/4,000 sq. m) colloidal phosphate
1 ton/acre greensand

For a soil initially of medium fertility:

15 tons/acre (13,610 kg/4,000 sq. m) manure or compost
¾ ton/acre (680 kg/4,000 sq. m) colloidal phosphate
¾ ton/acre greensand

For an initially fertile soil (just in case):

10 tons/acre (9,070 kg/4,000 sq. m) manure or compost
½ ton/acre (455 kg/4,000 sq. m) colloidal phosphate
½ ton/acre greensand

In all cases add sufficient limestone to maintain a pH of 6.5.

Maintenance Application

(applied in year 5)

½ ton/acre colloidal phosphate
½ ton/acre greensand
Limestone as required
    (not necessary if a soil test indicates that
    P, K, and trace minerals are adequate)

Maintenance Application

(applied every other year)

10–20 tons/acre (9,070 to 18,145 kg/
    4,000 sq. m) manure or compost

Note: I decided to leave these recommendations (slightly modified) in this revised edition because this is what I did when I started in 1965, and it worked. I still add some phosphate rock and greensand to new land, but I no longer purchase them for the land we have been cultivating for many years. Once a vigorous soil biology is established, we are able to maintain it with the very minimal input systems described in chapter 13. However, you can follow these general guidelines if you have no more specific recommendations from professional soil tests or based on a known history of previously applied soil amendments.

Where green manures or grazed sod crops figure prominently in the crop rotation, the quantities of manure or compost can be lowered as conditions indicate.

Agriculture, and in many experiment station publications. As Hopkins understood it:

The real question is, shall the farmer pay ten times as much as he ought to pay for food to enrich his soil? Shall he buy nitrogen at 45 to 50 cents a pound when the air above every acre contains 70 million pounds of free nitrogen? Shall he buy potassium at 5 to 20 cents a pound and apply 4 pounds per acre when his plowed soil already contains 30,000 pounds of potassium per acre, with still larger quantities in the subsoil? Because his soil needs phosphorus, shall he employ the fertilizer factory to make it soluble and then buy it at 12 to 30 cents a pound in an acid phosphate or "complete" fertilizer when he can get it for 3 cents a pound in the fine-ground natural rock phosphate, and when, by growing and plowing under plenty of clover (either directly or in manure), he can get nitrogen with profit from the air, liberate potassium from the inexhaustible supply in the soil, and make soluble the phosphorus in the natural rock phosphate which he can apply in abundance at low cost?[5]

Prices may have changed, but the basic truths about long-term soil fertility and economic success for the farmer are as clear now as they were then.

The efforts of Cyril Hopkins serve as a metaphor for independent truths set up against an advertising and sales blitz that tries to pretend the truths don't exist. The result of more than a century of fertilizer salesmanship is that no one today remembers Cyril Hopkins. The soil fertility truths that he champi-

oned, although they were understood for generations, have been forgotten for so long that they are now regarded as some sort of revolutionary heresy.

Hopkins was well aware of that possibility. He wrote numerous experiment station bulletins encouraging farmers to realize that no salesman was going to tell them about these ideas because there was so little to sell. He warned them that the large fertilizer manufacturers were concerned first and foremost with selling and only secondarily with farming. He predicted that the manufacturers would push their products endlessly, until farmers forgot how well agriculture could work with a bare minimum of purchased materials. Well, Cyril Hopkins may have lost that struggle and been momentarily forgotten, but the truth of "permanent soil fertility" is still right there in the earth for those who care to look.

Growing an Acre of Corn: Two Approaches

Sustainable: Feed the Soil

Soil fertility is understood as a biological process. Once established, fertility can be maintained and improved by crop rotations that include legumes plus the addition of mineral raw materials.

Only the actual quantity of nutrients that leave the farm in stock or crops sold need to be purchased as inputs to maintain fertility. Nitrogen is not a purchased input because it is supplied by symbiotic and non-symbiotic processes.

Inputs are purchased in their least processed and least expensive form. Nutrient solubility and availability is considered to be a natural function of the biological processes in a properly managed soil.

Any feed brought onto the farm is calculated on the plus side of the input ledger. On average, 75 percent of the nutrient value of feed consumed by animals is returned in the manure as a nutrient input to the farm.

Non-sustainable: Feed the Plant

Soil fertility is understood to be an imported commodity. It is supplied by fertilizer inputs from off the farm, which are calculated in terms of so many hundred pounds of fertilizer applied to "create" the crop.

All the nutrients (N, P, K, Ca, Mg) known to be required to "create" 1 acre of corn (roots-tops-grain) are purchased as inputs each year. Nitrogen is a very important purchased input.

Inputs are purchased in their most processed and most expensive form. Nutrient solubility and availability are considered to be industrial functions of the chemical processes in a fertilizer factory.

Any feed brought onto the farm is calculated solely as a feed expense and is not credited for its manure value. Animal manure, in general, is treated as a problem rather than as an asset.

CHAPTER THIRTEEN

Farm-Generated Fertility

Vegetable growers cultivate the soil intensively for fast-growing annual crops. Unlike the perennial culture of pastures, hay fields, and orchards (in which organic-matter levels are maintained because the soil is permanently covered with vegetation) or field crops such as corn and soybeans (in which a rotation with small grains and legumes can maintain production levels), the demands of intensive vegetable growing require extra inputs of organic matter. Those inputs have traditionally been acquired by importing animal manures from other farms. However, as those other farms begin to modify their production practices to create more sustainable, low-input systems, they are logically coming to consider the animal manures as valuable resources for their own use. One immediate result is that manure, when available, has risen considerably in price. Another change affecting manure supplies is the demise of so many farms. More often than not the neighboring farm no longer exists.

I encountered that latter reality in 1990 when I returned to my farm in Maine after 12 years running farms and research projects in other parts of the United States. Farming is not a growth industry in my area, and there was only one dairy farm left in the whole county. Four years later there were none. Fortunately, I had begun exploring the potential of farm-generated systems that didn't import manure many years earlier. I had always been bothered by my dependence on manure from other farms. It seemed to me like the one flaw in my organic vegetable-production system. I

could understand the use of imported manure in the early years, when it provided almost instant soil improvement; or on a small acreage with all of the land in vegetable production. But if organic vegetable growing continues to succeed only because there are nearby farms from which fertility can be robbed, then it clearly cannot be portrayed as either a universal or a sustainable means of producing food.

Even though I employ all the green manures and ingenious crop rotations I can devise, I still need extra organic matter for my intensive production. If I were located near a supply of municipal leaf compost, a horse racetrack, or some similar resource, I could investigate that option. But there are none nearby, and I am also concerned about the potential contamination of those products in a world where organic wastes are all too often polluted by toxins. The latest evidence on many of the municipal and waste product composts cites high levels of undesirable heavy metals. And as the authors of one study state bluntly, once heavy metals are introduced into the soil it is hardly possible to get them out.[1] The obvious solution is to grow my own raw materials and compost them.

My original inspiration came many years ago when I drove past a defunct neighboring farm that had been bought up as a vacation home by out-of-staters. They had hired a local contractor to rotary-mow the fields and pastures to maintain the "farm look." My first thought was why didn't I try to get hired to do that and also convince the owners that they should additionally pay me to collect and haul off all those unsightly grass wastes. Why continue to buy spoiled hay for mulching if I could get it free or at a profit?

As I pondered the delightful scenario of being paid to truck away raw materials for mulching, light bulbs began to appear in the comic-strip balloon above my head. Why not compost the chopped grass? What was manure anyway but hay and other forages chewed and digested by livestock? Isn't a rotary mower the equivalent of the cow's teeth? (I now call it the "iron cow.") Aren't the bacteria in soil and in compost heaps as effective at breaking down those raw materials as the digestive bacteria in the cow's stomach? Wouldn't turning the compost heap duplicate the additional mechanical processing of the cow's cud chewing? My answers were positive to all of the above.

The idea is not new. As far back as the middle of the 19th century, and with increasing frequency in the early part of the 20th century as horsepower gave way to internal combustion, many growers who had depended upon horse manure from the stables to fertilize their vegetable land became concerned about finding new sources of organic matter. They experimented with composting organic wastes, using the terms *vegetable manure* and *artificial manure* to describe composts made directly from plant material without first feeding it to an animal. A few books and pamphlets were published before the interest faded.[2] All too soon, artificial manure became confused with artificial fertilizer, and chemicals momentarily seemed to be the answer farmers were looking for.

Like mine, most farms have potential forage land available. If not, it should be possible to rent land in the vicinity. In addition to old hay fields, there is usually land either too steep, too rocky, or too wet for vegetables. It makes sense to establish selected forage crops on all these areas and harvest them for composting. I realize this won't quite approximate the cyclical flow of natural systems, but it does begin to make my organic vegetable farm more honestly sustainable. On the surface this solution seems very straightforward. But to turn good concept into good practice, a number of specifics must be determined and questions answered.

1. Which forage sources will be most successful in such a system?

Based on experience and criteria of expense, ease, and yield, I have narrowed down my choices. Exotic, high-yielding crops like comfrey and Jerusalem artichokes were tried and rejected because of the extra work required for establishment and management. The easiest option is using old hay fields that presently exist. There are no establishment costs, and the hay field supports a mix of plant species. I also recommend new seedings of alfalfa on well-drained land, reed canary grass on wet land, or other high-yielding forages. Alfalfa is an obvious first choice. It is a perennial, deep-rooted, drought-resistant, leguminous forage crop that can be grown successfully in most parts of the country. Even in my short summer, three cuttings are possible.

An arable rotation can be used on land that will rotate back into vegetables after a few years. One that I like begins with seeding winter rye after clearing vegetables in the fall. The rye is undersown to a biennial sweet clover early the following spring. I mow the rye for compost material in midsummer and let the sweet clover grow on through the second winter. The clover is mowed the second summer and followed by buckwheat. I mow that in fall before sowing a mix of rye and hairy vetch, which occupies the land over the third winter until it is tilled under for vegetables the next spring. The land is out of vegetable production for two growing seasons, during which time I mow off heavy yields of three different forage crops for composting and end with a green manure.

2. At what stage of growth should the forage crops be harvested?

My experience and the information from the books I consulted agree that forages used for composting should ideally be harvested at the same stage as they would be for consumption by livestock. The carbon:nitrogen ratio at that stage is within the recommended limits for successful composting. Since the ideal harvesttime may also be a busy time for vegetable growers, I have done trials where I harvested the forages whenever it fit into my schedule, even late into the fall when all the hay was stemmy. Some of that material required two years to become finished compost, but it still worked, and I was very pleased with the results.[3]

3. What is the most efficient method for harvesting and composting?

I'm sure it might be nice to have a forage chopper and transport wagons and a compost turner. But I have always found that not having access to "ideal" equipment makes me more inventive. When I began these trials, I had access to a small tractor with a rotary mower and a hay rake. They worked just fine. My basic method is to mow the field and let the material wilt before raking it into windrows. I collect the windrows by using the loader on the tractor as a buckrake. I start at one end of a windrow, open the bucket up so the blade is vertical, set the edge about an inch (2.5 centimeters) above the soil, and drive forward as quickly as I can, pushing an increasing mound of forage in front of me as I move it across the field. After depositing all the windrows at one side of the field, I push them into a compost windrow with the bucket and let the decomposition process begin.

There is very little friction between dry forage and the stubble beneath it. The buckrake (also called a sweep rake) is a series of 6- to 8-foot- (1.8 to 2.4 meter) long parallel wooden poles that are spaced 1 foot (30 centimeters) apart and that can skim over the surface with a backboard behind them to push the hay. The rake takes advantage of that slippery surface to move large amounts of forage. Since the tractor bucket worked fine as is, I didn't need to add

wooden fingers, but I could have, or I could have made a simple buckrake from sapling poles that could be pushed by any old vehicle.

On the smaller scale I now have a rotary mower for my walking tractor. I borrow an old horse-drawn dump rake from a neighbor and pull it with the walking tractor. I am currently making a small buckrake for the walking tractor, but I have used a jury rig on the front of an old truck with no problem. For those with a bent toward handwork, the scythe is a very efficient mowing tool. In the old days a good worker with a scythe could mow an acre a day. I prefer the continental European models with a straight snath (handle) and lightweight blade. The first lesson, as with any cutting tool, is to keep it sharp. The second lesson is to keep the blade parallel to the ground through each smooth, circular stroke. You cut obliquely across the face of the uncut grass, not directly at it, taking a small strip with each stroke.[4]

The best part of mowing and collecting forage for composting is the lack of concern about weather. Anyone who has ever struggled to make hay in an uncooperative climate will appreciate this new, relaxed attitude. Instead of worrying about the hay crop being rained on, I look forward to it. Instead of carefully setting the hay rake to just skim the surface, I intentionally set it too deep so it will kick soil and old thatch into the new mowings. All the practices that would be negatives if I were striving to make hay for feeding four-legged livestock become positives when I want the mowings to compost. My new "livestock" are the microorganisms in the compost heap, and they love wet and dirty hay. This kind of haymaking puts a smile on my face.

I compost in windrows. I pay attention to getting all the material thoroughly moist to prevent fire-fanging. I lay a perforated sprinkler hose on the windrow and run it as needed. I try to turn the mown hay compost with the tractor bucket at least twice if I have time. I keep the heaps covered, using a fabric (mentioned earlier) that is permeable to air and

moisture but still blocks sunlight and sheds rain. I don't plan on using the compost until it is one and a half to two years old.

In order to enhance the decomposition process and the quality of the resulting compost, I incorporate 1 to 2 percent by weight of a montmorillonite-type clay with the forage crops in the windrow. Montmorillonite is an expanding-lattice clay that has been determined to have both biotic and abiotic effects in aiding the conversion of organic matter into stable humus.[5] The resulting clay humus fraction that develops is very beneficial to soil fertility and plant growth. Montmorillonite (also sold as Wyoming bentonite) is mined for numerous industrial and agricultural uses. I have obtained my supplies either through the livestock feed industry, where it is used as a binder for pelleted feeds, or directly from wholesalers.

4. How can fertility be maintained on the forage stands?

Since we are growing the forage crops in order to transport their organic matter and nutrients elsewhere, we must plan to supplement the land on which they grow so as to maintain the system. I could treat them as I have treated forage fields in the past, tilling them up every four or more years and incorporating lime and other low-solubility mineral nutrients before reseeding. Or if I wished to make my system as locally based as possible, I could fertilize with the finely ground rock powder waste from local rock-crushing operations.

Many of the popular vegetables are not strong feeders on the less soluble minerals in rock powders. However, many of the vigorous forage crops are strong feeders. Therefore, the system, in its simplest form (as I suggested in the previous chapter), would be to select bulk-yielding forage crops with strong feeding power on less soluble nutrient sources, grow them on land fertilized with locally available waste rock minerals, and then harvest and compost them

for the vegetable land. Alfalfa, for example, is a very strong feeder on insoluble soil minerals and in addition is very deeply rooted.

5. What is the ratio of acreage of feeding land (the forages) to that of consuming land (the vegetables)?

This would depend on the quality of the soil, the intensity of the vegetable cropping, and to some extent on the crops themselves. I suspect a 1:1 ratio would be adequate in most conditions, but I can see it going as high as 3:1 or 4:1 to get a low-fertility soil up to speed. In a delightful old book (*Farming with Green Manures* by Dr. C. Harlan, published in 1883), the author describes working to a 3:1 ratio, but in a very efficient manner. He had a 20-acre field of clover, which he mowed three times during the summer. Each time he mowed it, he raked the clippings from 15 of the acres (6 hectares) onto the remaining 5 acres (2 hectares). By the end of the summer he had an enormous quantity of material on the favored 5 acres. He plowed and harrowed that section and grew vegetables there the next summer while he mowed and raked to concentrate the clippings on a second 5-acre parcel. That new section grew vegetables during the third year, while the first piece went back into clover. He continued to rotate the 5-acre parcels. If, instead of 20 acres of good land, he had had only 5 acres suitable for vegetables and 15 unsuited, it could have worked just as well, and those otherwise unproductive acres would have made a real contribution.

A comparison with manure use is instructive. If the ratio of feeding land to consuming land is calculated based on the number of livestock that can be fed off the forage of an average acre and the quantity of manure produced, the old-time market gardener's manure application of 50 tons per acre (45,360 kilograms/4,000 square meters) is equivalent to spreading the production of 8 to 10 acres (32,375 to 40,470 square meters) of land onto 1 acre. Obviously, heavy applications of organic matter will raise soil fertility more quickly at the start, but a 1:1 ratio combined with seasonal green manures should be more than adequate to maintain fertility on any intensively cropped soil.

Another option in the search for farm-generated inputs is to work cooperatively with other farms. I visited a large organic vegetable farm in France that followed a 12-year crop rotation. For 3 years out of the 12 the land was in alfalfa, and it spent 1 year in wheat undersown to clover. A neighboring dairy farm harvested the alfalfa and gave the vegetable farm cow manure in return. Another neighbor harvested the wheat, for which they were paid, but left the chopped straw to till under with the clover. It was a nice system and a very productive farm.

Permanent Soil Fertility

In my pursuit of a permanent soil fertility that can be maintained with fewer inputs, I have added clay (specifically, montmorillonite/bentonite types) directly to the soil in my greenhouses. I spread the clay on top of a layer of peat moss (for long-lasting organic matter) and till them in together. That practice was inspired by Michigan State research,[6] which indicated impressive improvements in yield, moisture retention, and nutrient availability on sandy soils amended with montmorillonite clay. Montmorillonite has the highest cation exchange capacity (CEC) of all mineral soil components. Investigations of soils that suppress soilborne plant diseases have found the common thread to be their content of montmorillonite clay. But best of all, I like the idea of making a permanent addition to the soil, one that will effect a long-term improvement rather than a short-term stimulus.

Adding clay to sandy soils was a common practice in Europe and, to a lesser extent, in the United States in centuries past. Henry Colman, who investigated the practice thoroughly in his 1846 book, *European Agriculture and Rural Economy*, refers to the soil improvement by claying as being of a "substantial and permanent character." Colman describes his

visits to different farmers who were spreading clay at rates between 25 and 100 tons (22,680 to 90,720 kilograms) to the acre (4,000 square meters). Every one of them attested to the benefits of the practice in terms of higher yields and the higher quality of the produce. On one farm belonging to the Duke of Bedford, clay had been spread on a total of 420 acres (1.7 square kilometer). The common technique was to distribute the material on grassland or plowed fields in the fall, let it become well pulverized by winter freezes, and then incorporate it with a harrow in spring. Those spreading clay at the lighter rates noticed additional improvements if the application was repeated after 20 years' time.

Since I have a preference for correcting causes rather than treating the symptoms of problems, I am very interested in permanently correcting the low initial fertility of my sandy soil by adding clay. I can justify the use of purchased montmorillonite (at 10 tons per acre [9,070 kilograms/4,000 square meter]) on my greenhouse soils because they are cropped intensively over a long season. But the cost would be prohibitive if I wanted to do my outdoor fields. The opportunity to spread clay on the outdoor fields came after we had dug an irrigation pond. When we were planning the pond, I knew I would have uses for the peatlike, muck soil from the top layer. I further assumed, and rightly so, that the gravelly underlayer would be perfect for building farm roads. However, I was pleasantly surprised when huge piles of a rich, blue marine clay appeared from the deeper levels. I emulated the good Duke of Bedford, albeit on a much smaller scale, spreading it on my fields and leaving it over the winter before incorporation. The difference in crop yield and quality was apparent right from the start, especially with the onion family, which had never appreciated our sandy soil, and it looks like a permanent improvement.

The peatlike material from the top of the pond site was put to use improving the soil for a new greenhouse. All organic growers understand the importance of organic matter. But I have long felt

that compost or manure can be used to excess in greenhouses. It's entirely possible to get too much of a good thing.

In greenhouses, the best way to improve soil quickly is to get soil structure and plant nutrition from two different sources. I like peat for structure. I think of it as a "low-octane" soil improver as far as plant food is concerned. To extend my fuel metaphor, peat is valuable for helping to fill the tank, without being so powerful as to burn out the valves. When I use peat as the organic ingredient to take care of the structural part of soil improvement, I can subsequently use a much lighter compost application for just the right amount of nutrition.

The peat I used on that greenhouse soil came from this farm. But I have also purchased bales of peat for soil improvement as well as for potting mixes. I do not share the anti-peat-moss sentiment I occasionally hear expressed. The anti-peat movement began in Europe where, because of population density, limited peat deposits, and centuries-long use of the resource, they are at the point where finding substitutes for peat makes sense. But the same is not the case in North America. Of the peat lands in North America, only 0.02 percent ($2/100$ of 1 percent) are being used for peat harvesting. On this continent peat is forming some 5 to 10 times faster than the rate at which we are using it. And even if we don't include bogs located so far north that their use would never be economical, peat is still a resource that is forming much faster than we are using it. To my mind that is the definition of a renewable resource.

Obviously, it behooves us to make sure that every natural resource is managed sustainably and that unique areas are protected. My investigations into the peat moss industry don't give me cause to worry. Just out of curiosity, though, I have explored locally available peat alternatives. The crumbly insides of well-rotted maple and birch tree trunks on the forest floor gave reliable results in potting mixtures.

CHAPTER FOURTEEN

The Self-Fed Farm

In his classic book *Root Development of Field Crops*, John Weaver wrote, "Covering the land with grass is nature's way of restoring to old, worn out soils the productivity and good tilth of virgin ones. Grass is a soil builder, a soil renewer, and a soil protector." Weaver was very aware that the plants that will sustain the soil fertility to feed mankind into the distant future are familiar species, but they aren't food crops for us. Rather, they are the perennial grasses, legumes, and numerous deep-rooting edible forbs growing in pastures that are rotationally grazed by livestock. They are the keys to maintaining soil fertility and preventing soil erosion because they create fertile soil with good structure for the human food crops that follow livestock in the rotation.

Wendell Berry made an astute comment on just this situation: "Once plants and animals were raised together on the same farm—which therefore neither

Our laying hens on range.

produced unmanageable surpluses of manure, to be wasted and to pollute the water supply, nor depended on such quantities of commercial fertilizer. The genius of American farm experts is very well demonstrated here: they can take a solution and divide it neatly into two problems." The *solution* is mixed farming. Mixed farming produces both grazed livestock and a rotation of annual crops on the same farm. The *two problems* are feedlots for livestock, which concentrate both the animals and all their manure, and monocrop farms with their consequent soil erosion and excess fertilizer runoff.

In England the classic system of mixed farming (both crops and livestock) was called ley farming. There are published references to the concept as early as the 1600s. It was well explained by Robert Elliot in his book *Agricultural Changes* (1898) and further defined by Sir George Stapledon in *Ley Farming* (1941). *Ley* is an old English word for a temporary pasture—one that was traditionally tilled up to grow a few years of grain or vegetable crops after two to four years as a grazed sod. I will let Stapledon put it in his own words:

The case for ley farming, then, rests securely on the value and cheapness of young grass as feed when compared with all other feeds and on the energy-potential of the sod as the foundation upon which to build sensible and crop-producing rotations. . . .

The essence of ley farming is to grow crops and grass; and to be at as much pains to use the sod to the best advantage as a manure and the foundation of fertility as to use the grass to the best advantage as a feed. . . .

A healthy sod has many of the characteristics of a well-made and well-rotted compost, and management should aim at accentuating these characteristics. . . .

The grass sod has an importance as a means to aiding maximum crop production (wheat and other cereals, sugar-beet, roots, kale, potatoes, etc.) at least as great as its grass-producing function.

Another old name for ley farming was *alternate husbandry*, because the fields alternate between building fertility during the sod phase and exploiting it during the annual crop years. Today's researchers refer to it as a sod-based rotation. Experience suggests that the fertility of land tilled up after four years in a rotationally grazed mixed-species sod is practically that of virgin soil again. The combination of the enormous quantity of plant fiber from the extensive roots of the perennial pasture plants, plus the minerals they extract from the deeper soil levels, plus the manure deposited by the grazing animals provides ideal growing conditions for the annual crops that follow the sod in addition to improved soil structure to protect against erosion. The fertility stored up during four years in grazed pasture was considered sufficient for up to four years of grain, bean, and fodder crops before being sown back down to pasture again.

Until 2012 we had been maintaining the fertility of our vegetable land with green manures and purchased manure compost to supplement the compost we make on the farm. Both the expense of the purchased compost and the fossil-fuel use for delivering that material convinced us to try another system. I'd been reading about the idea of ley farming for many years and had come to the conclusion that it could be the key to maintaining perpetual soil fertility. We don't have enough good vegetable soil to leave in pasture for three to four years, but I found research studies from the 1930s that determined a one- to two-year sod would be sufficient on podzolic soils like ours.[1]

So we instituted a modified vegetable growers' version of ley farming on our limited area of good vegetable land. Early in the spring every year we now sow half our vegetable land to a grass/legume pasture mix. Once established, this pastureland is grazed rotationally during the summer and fall by our laying hens. It would be nice to have cattle or sheep as our livestock, as the early ley-farming enthusiasts did, but we have chosen eggs from pastured laying hens as the ideal livestock product to complement a vegetable farm. The other half of our vegetable land, which received that same fertility-enhancing pasture/grazing treatment the previous year, is tilled up section by section as we plant crops to grow this year's vegetables. (We till under the sod a full three to four weeks before we seed or transplant a vegetable crop. In combination with our rotary tiller, we use a chisel plow to aid in continually deepening our soil base.) Every year half the vegetable land is gaining fertility and the other half is exploiting the fertility produced the year before. If we did wish to graze four-legged livestock, we would probably need to leave the land in sod for two years to gain enough fertility from the growth of the pasture plants. With grain-fed livestock like our laying hens, however, three-quarters of the mineral value of their organic layer feed is spread on the land directly through their manure, thus imparting additional fertility to a shorter-term sod. For more information on moveable chicken houses see "The Chickshaw" section in chapter 27, page 215.

The aim of this alternate husbandry is to establish a system for near-perpetual maintenance of soil fertility from within the farm, and it has done just that. The only other input for this system, in addition to sunlight and CO_2 for photosynthesis (which are universally available), is most likely just some limestone every four years. During the cropping year we may use alfalfa meal or a locally available crab meal for a temporary fertility boost for succession crops. We especially like the alfalfa meal because it is a product we could grow and dry ourselves if we wished to make our soil-fertility program even more independent . Since we employ movable greenhouses, the greenhouse soil itself can benefit from the ley-farming concept when we move around our small, wheeled layer houses so the chickens can graze one of the uncovered plots. In my experience the fibrous remains of the roots of grasses and legumes that have been tilled under create a superior soil fertility for vegetable crops over that achieved from an equal quantity of organic matter added as compost.

And as an additional advantage beyond using just green manures, the period in soil-improving grass/legume sod adds directly to the farm's income through the sale of livestock products.

CHAPTER FIFTEEN

Direct-Seeding

The aim of seeding is to place the vegetable seeds directly in the ground where they are to grow. The key tool for this process is a precision seeder. The perfect precision seeder will plant any size seed at any desired spacing and at the proper depth with reliable accuracy.

Since seeds come in so many different sizes and shapes, this is no easy task. Many seeders work well only with round seeds or pelleted seeds that have been made round by coating them with a claylike material. Although pelleted seeds are easier for many seeders to handle, they are also more expensive and are available in only a limited number of varieties. Frequently, the varieties of greatest value to the small-acreage grower (by virtue of flavor, tenderness, texture, storage, specialty market, and so forth) are not the commercial bulk-shipping varieties and hence are not available as pelleted seed. Sowing is often more successful if the seed can be bought "sized"—that is, separated by small increments into lots of identical dimensions. Sized-seed lots also germinate evenly and grow with equal vigor; with most crops, these qualities correlate with seed size. Thus, crop uniformity and harvest predictability are further returns from a sized-seed lot.

Although many vegetable producers direct-seed the majority of their crops, I recommend direct-seeding only those vegetables that are not practically or economically feasible to transplant. These are the taprooted crops (carrot, parsnip); the low-return-per-square-foot crops (corn, pumpkin); the easily drilled crops (pea, bean); and the fast-growing crops (radish, spinach). My reasons for broad-scale reliance on transplanting are outlined in the next chapter. One result of this preference is that it simplifies the direct-seeding system to fewer crops. The right choice of precision seeder is then the key to planting those crops as effectively as possible.

Desirable Features

A good one-row, hand-pushed (or -pulled) precision seeder has certain features:

It is easy to sow in a straight line. This is crucial for subsequent cultivation. Straight rows can be mechanically cultivated right up to the seedlings, saving a prodigious amount of hand-weeding.

It gives precise seed placement. Good seeds cost money, and waste is expensive. Ideally, there should be no thinning required. When seeds are dropped where they are to grow and at the optimal spacing for best growth, the result is higher-quality produce. Overly crowded plants grow poorly and are slower to mature.

It allows accurate depth adjustment. Depth of planting affects germination, emergence, and early growth. The adjustment and maintenance of the necessary depth must be dependable. It is incumbent on the grower to provide a smooth seedbed without hummocks and hollows if the best results are to be obtained.

It is easy to fill and empty. In a multicrop system many different seeds are involved. The seeder should be designed so that the seed hopper is easy to fill and easy to empty of excess seed. Changing from one size cup, plate, or belt to another should not involve extra tools or complicated procedures.

It is flexible and adaptable. For a wide range of seeds, there needs to be a wide range of adjustments possible for seed size, seed spacing, and depth of planting. There should be no problem in making or obtaining specially sized cups, plates, or belts for the grower's needs.

There is a visible seed level and seed drop. Nothing is more frustrating than to seed a crop and then find out that the seeder was not functioning correctly, or that the seed supply ran out partway through. In the best models of precision seeders, the operator can clearly see whether the seeds are dropping and how many are left.

It includes a dependable row marker. The next row is marked by an adjustable marker arm while the previous row is being seeded. It is important to keep the rows spaced evenly. When rows are arrow-straight and equidistant, between-row and in-row cultivations go much faster.

A one-row Jang Precision Clean Seeder, JP-1.

Using the Seeder

The Earthway seeder, a plate model, was the first one I depended on. The one flaw that I and other growers have noted is that the plates are not stiff enough. Some small, hard seeds can wedge themselves behind the plate and bend it out from the wall, thus inhibiting seed pickup and delivery. My solution, with the plate I use for clover green manures, for example, is to purchase a blank seed plate and redrill the holes at a blunter angle so there isn't as much slope to force the seeds behind the plate. It might also be possible to use superglue and attach some stiffening ribs to the plates.

Another difficulty may arise with some seeds and some weather conditions where an electrostatic charge

builds up and holds the seeds to the plate. I solved that problem by fabricating a small wooden crossbar with a toothbrush attached below it. It sits across the seed hopper and brushes the seeds off the plate and down the seed tube. The manufacturer suggests washing the seed plate in soapy water and leaving a little soapy residue on it. The residue will prevent static cling. They also recommend periodic careful washing of the plates and the seed hopper to maintain the smooth and efficient functioning of the machine.

It is good policy to calibrate the seeder before using it for all the different seeds and other adjustments involved. This is most easily done by measuring the circumference of the driving wheel (let's say it is 36 inches [92 centimeters]). Turn the wheel a number of times (say three) by hand, with the seed hopper filled. The number of seeds that drop out of the seed tube (say 54) are the number

The dependable, lightweight, four-row Pinpoint Seeder.

that would be planted in 3 (revolutions) times 36 inches (circumference), or 108 inches (275 centimeters) of linear travel; 108 inches divided by 54 seeds is a seeding rate of 1 seed every 2 inches (5 centimeters). If that is the desired seeding rate, then note the cup, plate, or belt number on the seed packet for future reference. If not, make an adjustment and go through the process again.

Another excellent, albeit more expensive, seeder is the Jang JP-1, which uses seed rollers rather than plates to pick up the seeds.

Fudge Factor

Now just because it is possible to place a single seed at the precise spacing you desire, that doesn't always mean it should be done. Germination percentages must be considered. Seeds are sold with the germination percentage printed on the package, which was determined by a controlled laboratory test. Field germination will usually be lower than the stated percentage. Allowing for a fudge factor of 50 to 100 percent is wise when planning seeding rates. For example, if you want to end up with a plant every 4 inches (10 centimeters), the seeder could be set to drop a seed every 2 inches (5 centimeters). Less than perfect germination will do some of the thinning for you. Further thinning is quickly done with a hoe, because the seeds will be evenly spaced and not in a thick row or clump.

Marking

When seeding, a string should be stretched tightly to guide the first row. All subsequent rows will be marked with the row-marker arm on the seeder.

Johnny's row markers attached to their mesh roller.

With the four-row and the six-row seeders, the edge of the first pass marks where the next pass should go. Be sure to aim the seeder as straight and true as you can with each pass. When you're hand-seeding, you'll need to reset the string or tape measure for each row. For larger areas, you can use an adjustable rolling marker or marker rake.

Hand-Seeding

Hand-seeding is used for the cucurbit family—cucumber, pumpkin, summer squash, and winter squash. These are relatively large seeds that can be easily picked up with the fingers. The quantities and areas to be planted are not excessive for hand-seeding. Under poorer soil conditions these crops will benefit from being planted on small, soil-covered mounds of compost or manure for an extra fertility boost.

CHAPTER SIXTEEN

Transplanting

Transplanting is the practice of starting seedlings in one place and setting them out in another. In this way large numbers of young seedlings can be grown in a small area under controlled cultural conditions before they are taken to the field. Whether the seed-starting facility is indoors under lights or outdoors in a cold frame, poly tunnel, or greenhouse, the space must be kept clean and well maintained. Since conditions in any covered area cannot be completely natural, I take precautions by carefully removing all plant debris in the fall and making sure the space is empty of growing plants by early winter, so as to allow the low temperatures to do their work in freezing out pests. I recommend the use of a thermostatically controlled bottom-heat propagation mat to maintain optimal germinating temperatures when starting warm-weather crops.[1]

Transplanting has traditionally been used for those crops (celery, lettuce, onion, and tomato) that regrow roots easily. These crops don't suffer much from being transplanted, although they obviously grow better the less their roots are disturbed.

Transplanting is also of value for many crops (cucumber, melon, and parsley) that are less tolerant of root disturbance, but it must be conducted in such a way that the plants hardly know they have been moved. The best transplant system is one that does not disturb the roots, is uncomplicated, can be mechanized, and is inexpensive.

Greenhouse to Field

Transplanting should be understood as three separate operations: starting, potting on, and setting out.

Starting involves its own three subdivisions—type of containment, soil mix, and controlled climate. The seeds are sown in some sort of prepared bed or container. The container usually holds a special soil mix or potting soil. This mix differs from garden soil by being compounded of extra organic matter and drainage material, so the seedlings will thrive despite the confined conditions. A controlled climate is provided by growing the plants in a greenhouse, hotbed, cold frame, or sheltered area to enhance early growing conditions for the young seedlings.

Potting on means transferring the seedlings from the initial container to a larger container with wider plant spacing. With soil blocks or plug systems, this isn't always necessary from a practical point of view, except with those crops that are grown for a longer time or to a larger size before being set out. Potting on is always valuable from the perspective of the highest

field provides an almost certain harvest. Transplanting is the most reliable method for obtaining a uniform stand of plants with a predictable harvest date.

Transplanting is reliable because the grower has better control over the production environment. The germination and emergence variables that can be so unpredictable in the field are more certain in the greenhouse. The crops are uniform because there are no gaps in the rows. No land is wasted from a thin stand due to faulty germination. Vigorous transplants set out at the ideal plant density for optimal yield have a very high rate of survival. The harvest is predictable because the greatest variability in plant growth occurs in the seedling stage. Once they are past that stage, an even maturity and a dependable harvest can be counted on.

It is far easier to lavish extra care on thousands of tiny seedlings in a small space in the greenhouse than over wide areas in the field. During the critical early period of growth, when ideal conditions can make such a difference, the grower can provide those conditions with less labor and expense in a concentrated area.

Cheating Weeds and the Weather

When crops are sown in the field, weeds can begin germinating at the same time or even before. Direct-seeded crops may also need to be thinned, and they must contend with in-row weeds while young. Transplant crops start out with a three- to four-week head start on any newly germinating weeds, because the soil can be tilled immediately prior to transplanting. Further, since transplants can be set out at the final spacing, they do not require thinning and are much easier to cultivate for the control of any in-row weeds that may appear.

Transplants can measurably increase production on the intensively managed small farm, because they provide extra time for maturing succession crops. This is done by starting the succession crop as transplants three to four weeks before the preceding

plant quality, however, since only the most vigorous of the numerous young seedlings are selected.

Setting out is the process of planting the young plants in the field or in the production greenhouse where they are to grow. The greater the efficiency with which this transfer can be accomplished, the more cost-effective transplanting becomes as a component of vegetable crop production.

A Sure Harvest

Transplants assure the grower of crops throughout the growing season at the times and in the quantities required. A seed sown in the field is a gamble, but a healthy three- to four-week-old transplant set out in the

crop is to be harvested. Immediately after harvest, the ground is cleared, the plants are set, and the new crop is off and growing as if it had been planted three or four weeks earlier (which, of course, it was). The result is the same as if the growing season had been extended by three to four weeks. Transplanting allows less land to be used more efficiently for greater production.

Earlier maturity is another obvious advantage to transplanting. Plants started ahead inside and set out when the weather permits have a head start and will mature sooner than those seeded directly. In many cool climates, tomatoes, melons, peppers, and others are only successful as transplanted crops.

Transplanting Methods

In earlier days vegetable growers relied heavily on bare-root transplants; seedlings were dug up from a special bed or outdoor field and transplanted with no attempt to retain a ball of soil around the roots. Uniform results and good survival rates are difficult to achieve with this method. Most of the fine root hairs that supply the plant with water are lost upon uprooting. This reduces the absorbing surface of the root system and markedly delays the reestablishment and subsequent growth of the plants. This "transplant shock" can be avoided by moving plants without disturbing their fragile root systems.

Many types of containers have been used to keep the root ball intact—clay pots, plastic pots, peat pots, wood or paper bands, wooden flats, plug flats, and others. The plants and soil are either removed from the container before planting or are planted outside, container and all, if the pot (peat pot, paper band) is decomposable. Unfortunately, most containers have disadvantages. Peat pots and paper bands often do not decompose as intended and inhibit root growth. They are also expensive.

Traditional wooden flats grow excellent seedlings, but some of the seedling roots must be cut when removing the plants.

Individual pots of any type are time consuming and awkward to handle in quantity. The plug-type trays that contain individual cells for each plant solve the handling problem by combining the individual units. But they share a problem common to all containers—

TABLE 16.1. Transplant Timing

Crop	Transplant Timing
Artichoke	10 weeks after sowing
Basil	3 weeks after sowing
Beans	2 weeks after sowing
Beets	3 weeks after sowing
Broccoli	3 weeks after sowing
Cabbage	3 weeks after sowing
Cauliflower	3 weeks after sowing
Celeriac	6 weeks after sowing
Celery	6 weeks after sowing
Corn	As soon as the first green shoot appears
Cucumber	3 weeks after sowing
Eggplant	10 weeks after sowing
Fennel	3 weeks after sowing
Kale	3 weeks after sowing
Leek	Once seedlings are over 10" (25 cm) tall
Lettuce	3 weeks after sowing
Melon	3 weeks after sowing
Onion	6 weeks after sowing
Parsley	7 to 8 weeks after sowing
Peas	2 weeks after sowing
Pepper	10 weeks after sowing or before first flowers open
Potato	3 weeks after beginning pre-sprouting
Radicchio	3 weeks after sowing
Scallion	3 to 4 weeks after sowing
Spinach	2 weeks after sowing
Tomato	5–6 weeks after sowing
Winter squash	3 weeks after sowing
Zucchini	3 weeks after sowing

root circling. The seedling roots grow to the wall of the container and then follow it around and around.

Plants whose roots have circled do not get started as quickly after they are put out in the field. An attempt has been made to improve on this situation by designing the tray cells with a hole in the bottom for air pruning and ridges in the cell walls.

Fortunately, there is another kind of "container" better than all of the above. That container—the soil block—is the subject of the next chapter.

CHAPTER SEVENTEEN

Soil Blocks

It is always satisfying to find a technique that is simpler, more effective, and less expensive than what existed before. For the production of transplants, the soil block meets those criteria. The Dutch have been developing this technique for over 100 years, but the human experience with growing plants in a cube of "soil" goes back 2,000 years or more. The story of how cubes of rich mud were used to grow seedlings by the Aztec horticulturalists of the chinampas of Xochimilco, Mexico, makes fascinating reading.[1] A related technique is the old market gardener's practice of using 4- to 5-inch (10 to 13 centimeter) cubes of partially decomposed inverted sod for growing melon and cucumber transplants.

How Soil Blocks Work

A soil block is pretty much what the name implies—a block made out of lightly compressed potting soil. It serves as both the container and the growing medium for a transplant seedling. The blocks are composed entirely of potting soil and have no walls as such. Because they are pressed out by a form rather than filled into a form, air spaces provide the

Basil seedlings in soil blocks.

walls. Instead of the roots circling as they do upon reaching the wall of a container, they fill the block to the edges and wait. The air spaces between the blocks and the slight wall glazing caused by the block form keep the roots from growing from one block to another. The edge roots remain poised for rapid outward growth. When transplanted to the field, the seedling quickly becomes established. If the plants are kept too long in the blocks, however, the roots do extend into neighboring blocks, so the plants should be transplanted before this happens.

Despite being no more than a cube of growing medium, a soil block is not fragile. When first made, it is bound together by the fibrous nature of the moist ingredients. Once seeded, the roots of the young plant quickly fill the block and ensure its stability even when handled roughly. Soil blocks are the answer for a farm-produced seedling system that costs no more than the "soil" of which it is composed.

Advantages

The best thing about the soil-block system is that everything that can be done in small pots, "paks," trays, or plugs can be done in blocks without the expense and bother of a container. Blocks can be made to accommodate any need. The block may have a small depression on the top in which a seed is planted, but blocks can also be made with a deep center hole in which to root cuttings. They can be made with a large hole in which to transplant seedlings. Or they can be made with a hole precisely the size of a smaller block, so seedlings started in a germination chamber in small blocks can be quickly transplanted onto larger blocks.

Blocks provide the modular advantages of plug trays without the problems and expense of a plug system. Blocks free the grower from the mountains of plastic containers that have become so ubiquitous of late in horticultural operations. European growers sell bedding plants in blocks to customers, who transport them in their own containers. There is no plastic pot expense to the grower, the customer, or the environment. In short, soil blocks constitute the best system I have yet found for growing seedlings.

The Soil-Block Maker

The key to this system is the tool for making soil blocks—the soil-block maker or blocker. Basically, it

Lettuce seedlings in 2-inch blocks.

is an ejection mold that forms self-contained cubes out of a growing medium. Both hand and machine models are available. For small-scale production, hand-operated models are perfectly adequate. Motorized block-making machines have a capacity of over 10,000 blocks per hour. But they are overscaled for the small vegetable farm.

There are two features to understand about the blocker in order to appreciate the versatility of soil blocks: the size of the block form and the size and shape of the center pin.

THE FORM

Forms are available to make ¾-inch (2 centimeter) blocks (the mini-blocker), 1½-inch (4 centimeter) blocks, 2-inch (5 centimeter) blocks, 3-inch (7.5 centimeter) blocks, and 4-inch (10 centimeter) blocks (the maxi-blocker). The block shape is cubic rather than tapered. Horticultural researchers have found a cubic shape to be superior to the tapered-plug shape for the root growth of seedlings.

Two factors influence choice of block size—the type of plant and the length of the intended growing period prior to transplanting. For example, a larger block would be used for early sowings or where planting outside is likely to be delayed. A smaller block would suffice for short-duration propagation in summer and fall. The mini-block is used only as a germination block for starting seedlings.

Obviously, the smaller the block, the less potting mix and greenhouse space is required (a 1½-inch [4 centimeter] block contains less than half the volume of a 2-inch [5 centimeter] block). But in choosing between block sizes, the larger of the two is usually the safer choice. Of course, if a smaller size block is used, the plants can always be held for a shorter time. Or as is common in European commercial blocking operations, the nutrient requirements of plants in blocks too small to maintain them can be supplemented with soluble nutrients. The need for such supplementary fertilization is an absolute requirement in plug-type systems, because each cell contains so much less soil than a block. The popular upside-down pyramid shape, for example, contains only one-third the soil volume of a cubic block of the same top dimension.

My preference is always for the larger block, first because I believe it is false economy to stint on the care of young plants. Their vigorous early growth is the foundation for later productivity and pest resistance. I am convinced that the pest problems I often

Blocks versus Plugs

The quality of the transplant seedlings you grow is the first step toward a successful harvest. I think the minimal soil volume and the enclosed root space of a plug system turns out inferior seedlings that are prone to poor growth and later pest problems. When I see transplanted seedlings struggling on farms I visit, they are always from plug trays. Yes, soil blocks use more potting soil, but the extra expense is more than worth it for the security of healthy transplants that grow with no problem.

If you have been using a 98-cell plug tray, I recommend trying the "Stand-up 35" blocker from Johnny's Selected Seeds. It is sized to fill a 1020 tray with 105 1⅛-inch cubic blocks. That combination of blocker and tray will give you a modular system similar to a plug tray but with higher-quality, healthier seedlings.

Because of the better growing conditions, seedlings reach transplant size faster in blocks than in plug trays. See table 16.1, "Transplant Timing," for average seed-to-transplant times for soil blocks.

Professional hand soil-block makers for different sizes of blocks.

see on other farms are a consequence of transplanting seedlings already stressed from having been grown in the confines of multiple-cell plug trays. Second, I prefer not to rely on soluble feeding when the total nutrient package can be enclosed in the block from the start. All that is necessary when using the right-sized block and soil mix is to water the seedlings.

Another factor justifying any extra volume of growing medium is the addition of organic matter to the soil. If lettuce is grown in 2-inch (5 centimeter) blocks and set out at a spacing of 12 by 12 inches (30 by 30 centimeters), the amount of organic material in the blocks is the equivalent of applying 5 tons of compost per acre (4,535 kilograms/4,000 square meters)! Since peat is more than twice as valuable as manure for increasing long-term organic matter in the soil, the blocks are actually worth double their weight in manure. Where succession crops are grown, the soil-improving material added from transplanting alone can be substantial.

THE PIN

The pin is the object mounted in the center of the top press-form plate. The standard seed pin is a small button that makes an indentation for the seed in the top of the soil block. This pin is suitable for crops with seeds the size of lettuce, cabbage, onion, or tomato. Other pin types are dowel- or cube-shaped. I use the cubic pin for melon, squash, corn, peas, beans, and any other seeds of those dimensions. A long dowel pin is used to make a deeper hole into which cuttings can be inserted. Cubic pins are also used so a seedling in a smaller block can be potted on to a larger block; the pin makes a cubic hole in the top of the block into which the smaller block is placed. The different types of pins are easily interchangeable.

Blocking Systems

The ¾-inch (2 centimeter) block made with the mini-blocker is used for starting seeds. With this small block, enormous quantities of modular seedlings can be germinated on a heating pad or in a germination chamber. This is especially useful for seeds that take a long time to germinate, because a minimum of space is used in the process.

Mini-blocks are effective because they can be handled as soon as you want to pot on the seedlings. The oft-repeated admonition to wait until the first true leaves appear before transplanting is wrong. Specific investigations by W. J. C. Lawrence, one of the early potting-soil researchers, have shown that

the sooner young seedlings are potted on, the better is their eventual growth.[2]

The 1½-inch (4 centimeter) block is used for short-duration transplants of standard crops like lettuce, brassicas, beets, fennel, et cetera. When fitted with a long dowel pin, it makes an excellent block for rooting cuttings.

The 2-inch (5 centimeter) block is the standard for longer-duration transplants. When fitted with the ¾-inch cubic pin, it is used for germinating bean, pea, corn, or squash seeds and for the initial potting on of crops started in mini-blocks.

The 3-inch (7.5 centimeter) block fitted with a ¾-inch (2 centimeter) cubic pin offers the option to germinate many different field crops (squash, cucumber, melon) when greenhouse space is not critical. It is also an ideal size for potting on asparagus seedlings started in mini-blocks.

The 4-inch (10 centimeter) block fitted with a 1½- or 2-inch (4 or 5 centimeter) cubic pin can be the final home of artichoke, eggplant, pepper, and tomato seedlings. Because of its cubic shape, it has the same soil volume as a 6-inch (15 centimeter) pot. I now prefer 6-inch pots for these crops and I no longer use the 4-inch block.

Other Pin Options

In addition to the pins supplied with the blocker, you can make a pin of any desired size or shape. Most hard materials (wood, metal, or plastic) are suitable, as long as the pins have a smooth surface. You can use plug trays as molds, filling them with quick-hardening water putty to make many different sizes of pins that allow the integration of the plug and block systems.

Blocking Mixes

When transplants are grown, whether in blocks or pots, their rooting area is limited. Therefore the soil in which they grow must be specially formulated to compensate for these restricted conditions. For soil blocks, this special growing medium is called a blocking mix. The composition of a blocking mix differs from ordinary potting soil because of the unique requirements of block making. A blocking mix needs extra fibrous material to withstand being watered to a paste consistency and then formed into blocks. Unmodified garden soil treated this way would become hard and impenetrable. A blocking mix also needs good water-holding ability, because the blocks are not enclosed by a container. The bulk ingredients for blocking mixes are peat, sand, soil, and compost. Store-bought mixes can also work, but most will contain chemical additives not allowed by many organic certification programs. If you can find a commercial peat-and-perlite mix with no additives, you can supplement it with the soil, compost, and extra ingredients described below.

In the past few years, commercial, preformulated organic mixes with reasonably good growth potential have begun to appear on the market. However, shipping costs can be expensive if you live far away from the supplier. To be honest, I have found few of these products that will grow as nice seedlings as my own homemade mixes.

PEAT

Peat is a partly decayed, moisture-absorbing plant residue found in bogs and swamps. It provides the fiber and extra organic matter in a mix. All peats are not created equal, however, and quality can vary greatly. I recommend using the premium grade. Poor-quality peat contains a lot of sticks and is very dusty. The better-quality peats have more fiber and structure. Keep asking and searching your local garden suppliers until you can find good-quality peat moss. Very often a large greenhouse operation that makes its own mix will have access to a good product. The peat gives body to a block.

SAND

Sand or some similar granular substance is useful to open up the mix and provide more air porosity. A coarse sand with particles having a 1/16- to 1/8-inch (1.6 to 3.2 millimeter) diameter is the most effective. I prefer not to use vermiculite, as many commercial mixes do, because it is too light and tends to be crushed in the block-making process. If I want a lighter-weight mix I replace the sand with coarse perlite. Whatever the coarse product involved, adequate aeration is the key to successful plant growth in any medium.

COMPOST AND SOIL

Although most modern growing mediums no longer include any real soil, I have found both soil and compost to be important for plant growth in a mix. Together they replace the "loam" of the successful old-time potting mixtures.[3] In combination with the other ingredients, they provide stable, sustained-release nutrition to the plants. I suspect the most valuable contribution of the soil may be to moderate any excess nutrients in the compost, thus giving more consistent results. Whatever the reason, with soil and compost included there is no need for supplemental feeding.

Compost is the most important ingredient. It is best taken from two-year-old heaps that are fine in texture and well decomposed. The compost heap must be carefully prepared for future use in potting soil. I use no animal manure containing wood shavings or sawdust in the potting-mix compost and prefer horse manure if I can get it. I construct the heap with 2- to 6-inch (5 to 15 centimeter) layers of mixed garden wastes (such as outer leaves, pea vines, weeds) covered with a sprinkling of topsoil and 2 to 3 inches (5 to 7.5 centimeters) of strawy manure sprinkled with montmorillonite clay. Repeat the sequence until the heap is complete. The heap should be turned once the temperature rises and begins to decline so as to stimulate further decomposition.

There are no worms involved in our composting except those naturally present, which is usually a considerable number since we make compost on the same field every year. (I have purchased commercial worm composts, or castings, as a trial ingredient, and they did make an adequate substitute for our compost.) Both during breakdown and afterward the heap should be covered with a compost fabric. I strongly suggest letting the compost sit for an additional year (so that it is one and a half to two years old before use); the resulting compost is well worth the trouble. The better the compost ingredient, the better the growth of the plants will be. The exceptional quality of seedlings grown in this mix is reason enough to take special care when making a compost. Compost for blocking mixes must be stockpiled the fall before and stored where it won't freeze. Its value as a mix ingredient seems to be enhanced by mellowing in storage over the winter.

Soil refers to a fertile garden soil that is also stockpiled ahead of time. I collect it in the fall from land off which onions have just been harvested. I have found that seedlings (onions included) seem to grow best when the soil in the blocking mix has grown onions. I suspect there is some biological effect at work here, since crop-rotation studies have found onions (and leeks) to be highly beneficial preceding crops in a vegetable rotation.[4] The soil and compost should be sifted through a 1/2-inch (1.25 centimeter) mesh screen to remove sticks, stones, and lumps. The compost and peat for the extra-fine mix used either for mini-blocks or for the propagation of tiny flower seeds are sifted through a 1/4-inch (0.6 centimeter) mesh.

EXTRA INGREDIENTS

Lime, blood meal, colloidal phosphate, and greensand are added in smaller quantities.

Lime. Ground limestone is added to adjust the pH of the blocking mix. The quantity of lime is determined by the amount of peat, the most acidic ingredient. The pH of compost or garden soil should not need modification. My experience, as well as recent research results, has led me to aim for a growing-medium pH between 6 and 6.5 for all the major transplant crops. Those growers using different peats in the mix may want to run a few pH tests to be certain. However, the quantity of lime given in the formula below works for the different peats that I have encountered.

Blood meal. I find this to be the most consistently dependable slow-release source of nitrogen for growing mediums. English gardening books often refer to hoof-and-horn meal, which is similar. I have also used crab-shell meal with great success. Recent independent research confirms my experience and suggests that cottonseed meal and dried whey sludge also work well.[5]

Colloidal phosphate. A clay material associated with phosphate rock deposits and containing 22 percent P_2O_5. The finer the particles the better.

Greensand (glauconite). Greensand contains some potassium but is used here principally as a broad-spectrum source of micronutrients. A dried seaweed product like kelp meal can serve the same purpose, but I have achieved more consistent results with greensand.

The last three supplementary ingredients—blood meal, colloidal phosphate, and greensand—when mixed together in equal parts are referred to as the base fertilizer.

BLOCKING MIX RECIPE

A standard 10-quart bucket is the unit of measurement for the bulk ingredients. A standard cup measure is used for the supplementary ingredients. This recipe makes approximately 2 bushels of mix. Follow the steps in the order given.

3 buckets brown peat
½ cup (120 ml) lime
2 buckets coarse sand or perlite
3 cups (720 ml) base fertilizer
1 bucket soil
2 buckets compost

First, combine the peat with the lime, because that is the most acidic ingredient. Then add the sand or perlite. Mix in the base fertilizer next. By incorporating the dry supplemental ingredients with the peat in this manner, they will be distributed as uniformly as possible throughout the medium. Add the soil and compost, and mix completely a final time.

To use this recipe for larger quantities, think of it measured in "units." The unit can be any size, so long as the ratio between the bulk and the supplementary ingredients is maintained. A unit formula would call for:

30 units brown peat
⅛ unit lime
20 units coarse sand or perlite
¾ unit base fertilizer
10 units soil
20 units compost

MINI-BLOCK RECIPE

A different blend is used for germinating seeds in mini-blocks. Seeds germinate better in a "low-octane" mix, without any blood meal added. The peat and compost are finely screened through a ¼-inch (0.6 centimeter) mesh before being added to the mix.

16 units or 4 gallons (15 l) brown peat
⅛ unit or ½ cup (120 ml) colloidal phosphate
⅛ unit or ½ cup (120 ml) greensand
4 units or 1 gallon (3.8 l) well-decomposed compost

Note: If greensand is unavailable, leave it out; do not substitute a dried seaweed product in this mix.

STERILIZING THE MIX

In more than 20 years of using homemade mixes, I have never sterilized them. And I have not had problems. I realized early on that damping-off and similar seedling problems, which are usually blamed on unsterilized soil, are actually a function of cultural mistakes like overwatering, a lack of air movement, not enough sun, overfertilization, and so forth. Good, fertile garden soil and well-prepared compost contain many organisms that benefit seedling growth. If you "sterilize" these ingredients, you lose the benefits of a live mix without gaining the advantages that are achieved through proper seedling management. Recent university studies agree and emphasize the specific value of finished compost as a disease-suppressing ingredient in growing mixes.[6]

NITROGEN REACTION

With certain crops (mostly the more delicate bedding-plant flowers) there may be a further consideration. Where organic sources of nitrogen like blood meal or the old-time hoof-and-horn meal are included in a mix, the mineralization of the nitrogen by biological processes and the consequent production of ammonia can inhibit plant growth for a period of time after the mix is made, especially if moisture and temperature levels are high.[7] If you use a dried seaweed product instead of greensand, this consideration probably applies as well. To avoid this reaction, make up the mix fresh as you need it and never store it for more than three weeks. To my knowledge I have never been bothered by this problem, but I feel it is worth mentioning. In my experience, when the mix is stored for more than three months, it actually gets better, as all the ingredients seem to mellow together.

One of the European organic farms I visited actually processed their mix for a whole year before using it by layering the ingredients—horse manure on the bottom, then leaf mold, then compost—in a cold frame and growing first a crop of cabbage, followed by melons, then mâche. After the manure, leaf mold, and compost had been "processed" for a year by the roots of those crops, they became the basis for the mix. Seven parts of the processed ingredients were mixed with three parts of peat, rock powders were added, and the mix was ready. Another grower of my acquaintance uses only pure compost for growing seedlings in flats and plug trays. I relate those stories

as examples of the extremely wide variety of answers that different growers have found to the potting-soil question. The formulas I have given above are the answers that have worked well so far for me. They are not the only answers. If you prefer to purchase an organic mix rather than make your own, I can heartily recommend Vermont Compost Company. Their product, called "Fort V," makes exceptional soil blocks and grows exceptional seedlings.

MOISTENING THE MIX

Water must be added to wet the mix to blocking consistency. The amount of water varies depending on the initial moisture content of the ingredients. On average, to achieve a consistency wet enough for proper block making, the ratio of water to mix by volume will be about 1 part water to every 3 parts mix. A little over 2½ gallons (9.5 liters) of water should be added to every cubic foot (0.09 square meter) of mix.

For successful block making, be sure to use a mix that is wet enough. Since this will be much wetter than potting mixes used for pots or flats, it takes some getting used to. The most common mistake in block making is to try to make blocks from a mix that is too dry. The need to thoroughly moisten the mix is the reason the mix requires a high percentage of peat, to give it the necessary resiliency.

Handling Soil Blocks

Many large block-making operations set the newly formed and seeded blocks by the thousands on the concrete floor of the greenhouse. When they are ready to go to the field, the blocked seedlings are lifted with a broad, fine-tined fork and slid into transport crates. These crates have high sides so they can be stacked for transport without crushing the seedlings. In lieu of these special crates, three other options are practical for small-scale production.

Years ago we built simple three-sided wooden flats. The inside dimensions are 18¾ inches long by 8 inches wide by 2 inches high (48 centimeters by 20 centimeters by 5 centimeters). Three-quarter-inch (2 centimeter) stock is used for the sides and ½-inch (1.25 centimeter) stock for the bottom. One flat holds 60 of the 1½-inch (4 centimeter) blocks, 36 of the 2-inch (5 centimeter) blocks, or 18 of the 3-inch size (7.5 centimeter). These block flats are efficient to use in the greenhouse, because the benches need to be no more than 2×4s spaced to hold two rows of flats side by side. Low-sided flats such as these are not stackable when filled with plants. For transport, a carrying rack with spaced shelves is required.

The flats have only three sides so the blocked seedlings can be easily removed from the open side one at a time as they are being transplanted in the field. The flat is held in one hand by the long side while blocks are quickly placed in holes in the soil with the other. Similar three-sided flats (half as wide and only ¾ inch [2 centimeters] high at the sides) are used for mini-blocks. Since they are the same length as the others, they fit two to a space on the greenhouse bench for modular efficiency. Each of these flats holds 120 mini-blocks.

POLYCARBONATE FLATS

Our present very simple soil block flats came about after we had covered the end walls of a couple of new greenhouses with polycarbonate sheets. We cut the scraps left over into 8-by-18-inch (20 by 45 centimeter) pieces that hold 60, 36, or 18 blocks the same as the wooden trays. The polycarbonate flats last many years since they don't decompose like the wooden flats did.

One recent development has been to make block makers that fit the dimensions of the standard 10-20 trays that are so ubiquitous in American greenhouses. Many growers, familiar with the 10-20 trays and wishing to move beyond plug seedlings, have adopted this option.

BREAD TRAYS

When handling greater quantities of blocks, you can use the large plastic-mesh bread trays seen in bread delivery trucks. They can generally be bought used at a reasonable cost from regional bakeries. Since the sides on these trays are higher than all except the tallest seedlings, they can be stacked for transport. Bread trays vary in size, but on average each tray can hold 200 of the 1½-inch (4 centimeter) blocks and proportionally fewer of the larger sizes.

Results are excellent with bread trays. What with the open-mesh sides and bottom plus the air spaces between the blocks, the roots of the seedlings remain poised at all five potential soil-contact surfaces. The bread trays are not as easy to handle for field transplanting as the smaller three-sided flats, but they become manageable with practice.

Making Soil Blocks

Spread the wet mix on a hard surface at a depth thicker than the blocks to be made. Fill the soil-block maker by pressing it into the mix with a quick push and a twisting motion to seat the material. Lift the blocker, scrape off any excess mix against the edge of a board, and place the blocker on the three-sided flat, the bread tray, the plastic sheet, a concrete floor, or another surface. The blocks are ejected by pressing on the spring-loaded handle until a little moisture oozes out and then raising the sides of the form in a smooth, even motion.[8] After each use dip the blocker in water to rinse it. A surprising rate of block production (one grower claims up to 5,000 per hour using the 1½-inch [4 centimeter] commercial-scale model) will result with practice.

SEEDING THE BLOCKS

Each block is formed with an indentation in the top to receive the seed. The handmade blocks are usually sown by hand. With the motorized blockers, the sowing as well as the block forming is mechanized. An automatic seeder mounted over the block belt drops one seed into each indentation as the blocks pass under it. These motorized models are too large and expensive for the small-scale grower, but if a group of growers get together, there is a role for one of them in a specialized seedling operation. Small farmers always benefit from such cooperative arrangements and should consider participating whenever the opportunity arises.

SINGLE-PLANT BLOCKS

Sow one seed per block. There is a temptation to use two (just to be on the safe side), but that is not necessary. Germination is excellent in soil blocks because of the ease with which ideal moisture and temperature conditions can be maintained. The few seeds that don't germinate are much less of a problem than the labor to thin all those that do. Of course, if the seed is of questionable vitality, it is worth planting more than one seed per block, but obviously it pays to get good seed to begin with.

Seeding can be done with the fingers for large seeds such as cucumber, melon, and squash. Finger-seeding is also possible for small seeds that have been pelleted, although pelleted seeds are not easily available in most varieties, and naked seeds are more commonly used. The small seeds can be most accurately handled by using a small thin stick, a sharpened dowel, a toothpick, or a similar pointed implement. Spread the seeds on a dish. Moisten the tip of the stick in water and touch it to one seed. The seed adheres to the tip and is moved to the seed indentation in the top of a block and deposited there. The solid, moist block has more friction than the tip of the stick, so the seed stays on the block.

Another obvious technique is to crease one side of a seed packet or use any other V-shaped container and tap out the seeds by striking the container with your fingers or a small stick. The Park Seed Company sells small seeds in packets made of a heavy metal foil. If you take a pair of scissors and cut and crease an empty packet, the resulting "seed tapper" works exceptionally well, even for tiny seeds. Put only enough seeds in the packet at one time so they can be tapped out in a single row without bunching up.

Commercial seeding aids are available that aim to either wiggle, click, or vibrate the seeds out one by one. There are electrically operated vacuum seeders for the small-scale grower that can be adapted to seeding soil blocks. I have experimented with a non-electric, homemade vacuum seeder specifically for mini-blocks, which gets its suction from the return stroke of a foot-powered pump for rubber rafts. I haven't quite perfected the suction tips yet, but I will get it right one of these days. Instructions for a homemade, multipoint vacuum seeder are given in a past issue of *HortScience*.[9] Growers should try such aids and decide for themselves whether they are worth it. Whichever method you use, though, seed carefully to ensure that the seeds are accurately planted in each block.

In practice, these planting techniques quickly become efficient and precise. Remember that for many crops the soil-block system avoids all intermediate potting on. Crops are started in the block and later go directly to the field. That savings in time alone is worth the effort required to become proficient at single-seeding.

Germination

I never cover the seeds planted in mini-blocks. Oxygen is important for high-percentage seed germination. Thus, even a thin covering of soil or potting mix can lower the germination percentage. I find that to be important for all small flower seeds also. If the sowing instructions suggest the seeds need darkness to germinate, I cover the flats temporarily with a sheet of black plastic. I keep the moisture level high during the germination period by misting frequently with a fine spray of water. For the majority of crops in the larger blocks, I get sturdier seedlings if I cover the seeds. I do that by sprinkling a thin layer of potting soil over the top of the blocks.

The third key to a high germination percentage, in addition to air and moisture, is temperature. Ideal temperature for germination can best be maintained by using a thermostatically controlled soil-heating pad under the blocks.[10] The temperature is controlled at the desired setting by a remote thermostatic probe inserted into the potting soil or in the gap between the soil blocks. I use a temperature of 70 to 75°F (21–24°C) for most crops. For asparagus, cucumber, tomato, eggplant, melon, pepper, and squash, I use a setting of 80 to 85°F (27–30°C).

Multiplant Blocks

Although I have stressed the wisdom of sowing only one seed per block, there is an important exception to that rule—the multiplant block. In this case 3 to 12 seeds are deliberately planted in each block with no intention of thinning. Many crops grow normally under multiplant conditions, and transplant efficiency is enhanced by putting out clumps rather than single plants.

The concept of the multiplant block is based on spatial rather than linear plant distance in the field. For example, say the average ideal in onion spacing is one plant every 3 inches (7.5 centimeters) in rows spaced 12 inches (30 centimeters) apart. Multiplants aim at an equivalent spacing of four onions per square foot (0.09 square meter). The difference is that all four onions are started together in one block and grow together until harvest. Since it is just as easy to grow four plants to the block as it is to grow one, there is now only a quarter the block-making work and greenhouse space involved in raising the same number of plants. A similar advantage is realized when transplanting the seedlings to the field. When four plants can be handled as one, then only a quarter as many units need to be set out. Although bunched together, the plants will have extra space all around them. The onions grow normally in the

A well-set-up seedling house.

clump, gently pushing one another aside, attaining a nice round bulb shape and good size.

Not only bulb onions but also scallions (green onions) thrive in multiple plantings. Scallions are seeded 10 to 12 per block and grow in a bunch ready to tie for harvest. Weeding between the plants in the row is no longer a chore, since the wider spaces allow for easy cross-cultivation with a hoe. Obviously, multiplant blocks must be transplanted to the field a bit sooner (at a younger age) than single-plant blocks because of the extra seedling competition in the limited confines of the block.

Multiplant blocks can be sown either seed by seed or in bunches. For counting out seeds, the wiggle, click, and vibration seeders have a place here in speeding up the seeding operation, though at the sacrifice of some accuracy. When I need to be precise, I first tap the number of seeds required into a ¼-teaspoon measure so I can be sure of the count before sowing the block. Tiny scoops or spoons or other small-volume measures can be fabricated by the grower (or purchased from a kitchen-supply store) that will hold 5, 12, or whatever number of seeds. These are used to scoop up the seeds and dump them in each seed indentation. This method is not as accurate as counting, but it is a lot faster.

Multiplant blocks are an efficient option for a number of crops. In my experience onions, scallions, beets, parsley, spinach, corn, pole beans, and peas have been outstandingly successful in multiple plantings. Spinach, corn, pole beans, and peas, which are rarely transplanted, even for the earliest crop, become a much more reasonable proposition when the transplant work can be cut by 75 percent. European growers claim additional good results with cabbage, broccoli, and turnips planted at three to four seeds per block.

Watering

Blocks are made in a moist condition and need to be kept that way. Their inherent moistness is what

TABLE 17.1. Instructions for Transplants

Crop	Starting Block (inches)	Number of Seeds	Temperature	Pot on to*
Artichoke	1½	1 ★	70°F (21°C)	6" pot
Basil	¾	1	85°F (30°C)	2" block
Beans	2 ◆	1 ★	70°F (21°C)	
Beets	2	2 ★	70°F (21°C)	
Broccoli	2	1 ★	70°F (21°C)	
Cabbage	2	1 ★	70°F (21°C)	
Cauliflower	2	1 ★	70°F (21°C)	
Celeriac	¾	1	70°F (21°C)	2" block
Celery	¾	1	70°F (21°C)	2" block
Corn	3 ◆	3 ★	85°F (30°C)	
Cucumber	3 ◆	1 ★	85°F (30°C)	4" pot at 10 days
Eggplant	¾	1	85°F (30°C)	2" block ▶6" pot
Fennel	1½	1 ★	70°F (21°C)	
Kale	1½	1 ★	70°F (21°C)	
Lettuce	¾	1	70°F (21°C)	2" block
Melon	3 ◆	1 ★	85°F (30°C)	4" pot at 10 days
Onion	1½	4 ★	70°F (21°C)	
Parsley	¾	1	70°F (21°C)	2" block
Peas	2 ◆	2 ★	70°F (21°C)	
Pepper	¾	1	85°F (30°C)	2" block ▶6" pot
Radicchio	1½	1 ★	70°F (21°C)	
Scallion	1½	12 ★	70°F (21°C)	
Spinach	1½	3 ★	70°F (21°C)	
Tomato	¾	1	85°F (30°C)	2" block ▶6" pot
Winter squash	3 ◆	2 ★	85°F (30°C)	4" pot at 10 days
Zucchini	3 ◆	1 ★	85°F (30°C)	4" pot at 10 days

◆ = Cubic hole filled with seed starting mix

★ = Seeds covered with a thin layer of potting soil

▶ = Pot on to

* We use 4- and 6-inch (10 and 15 centimeter) square pots. For some crops we transplant seedlings from the seed block to a 2-inch (5 centimeter) block, then a 6-inch square pot.

makes them such an ideal germination medium. It is therefore most important that blocks are not allowed to dry out, which can result in both a check to plant growth and difficulty in rewetting. When blocks are set out on a bench or greenhouse floor, the edge blocks are the ones that are most susceptible to drying. A board the same height as the blocks placed along an exposed edge will help prevent this. Since the block has no restricting sides, the plants never sit in too much water. The block itself will take up no more water than it can hold.

To prevent erosion of the block, watering at first should be done gently with a very fine rose. If the rose is not fine enough, the mini-blocks should be misted rather than watered. Once the plants in blocks are growing, you can apply water through any fine sprinkler. Extra care in attention to watering is a general rule in successful block culture. It will be repaid many times over in the performance of the seedlings.

Soil blocks grow great root systems with no circling.

Potting On

Potting on is the practice of starting seeds in smaller blocks and then setting those blocks into larger blocks for further growth. Since most crops benefit from bottom heat to ensure and speed up germination, this practice makes efficient use of limited space in germination chambers or on heating pads. For example, 240 mini-blocks fit into the same space as only 36 of the 2-inch (5 centimeter) blocks.

Potting on blocks is quickly accomplished in a third the time required for potting on bare-root seedlings. The smaller block easily fits into a matching-sized hole in a larger block. The mini-blocks are usually potted on to 2-inch blocks and those, in turn, to 6-inch (15 centimeter) pots.

The 2-inch blocks are easily potted on using your fingers. For the mini-blocks, some form of transplant tool for lifting the blocks and pressing them into the cavity will be useful. One of the best implements for this job is a flexible artist's palette knife. It provides the extra dexterity necessary to handle mini-blocks with speed and efficiency.

Keep Them Growing

Potting on should be carried out as soon as the seeds have germinated in the mini-blocks and before the roots begin growing out of the small cubes. The less stress seedlings encounter, the better. Crops like tomatoes and peppers need to be given progressively more space as they grow. They produce the most compact transplants when they are spaced far enough apart so that their leaves never overlap those of another plant.

CHAPTER EIGHTEEN

Setting Out Transplants

Moisture is the first concern when setting out transplants. Soil-block plants should be watered thoroughly before being put into the ground. At first the amount of moisture in the block is more important to the establishment of the plant than the moisture level of the surrounding soil. The moisture level of the block allows the plant to send out new roots into the soil. Only after roots are established does the soil moisture become more important. Blocks should be very wet at the start and should be kept moist during the transplant operation. The carrying flats and transport rack should be shaded from the sun and shielded from drying winds.

The second concern is soil contact. The transplanted blocks must be placed lightly but firmly into the soil. Avoid air pockets and uncovered edges. If

Newly transplanted leek seedlings. When mature they will be protected by the moveable greenhouse in winter.

transplanting is to deliver all the benefits we've discussed, it must be done well. I recommend irrigating immediately following transplanting, and not only to provide moisture. The action of the water droplets also helps to cover any carelessness when firming the plants in. Although the wet block planted into dry soil will support itself surprisingly well, it can eventually suffer from stress. Irrigation is stress insurance.

Consistent depth of setting is also important for rapid plant establishment, even growth, and uniform maturity. The blocks should be set to their full depth in the soil. If a corner is exposed to the air, the peat in the blocks can dry out quickly on a hot, sunny day and set the plant back. On the smallest scale, transplant holes are made with a trowel. There are a number of designs for soil-block trowels that are easier on your wrist than a standard model: the dagger style and the right-angle trowel. Both are jabbed

into the soil and pulled back toward the operator to make a neat hole for setting the plant.

I make my own dagger-style model using a bricklayer's trowel with a 2-by-5-inch (5 by 13 centimeter) blade. I first cut off 2¼ to 2½ inches (6 to 6.5 centimeters) to shorten the blade, then bend the handle down to below horizontal at about the same angle that it was above. I now have a very efficient transplant trowel for soil blocks. The same tool can be used like a spatula to lift blocks from the flat.

When you're setting out plants, be sure to space them correctly. Accurate spacing not only makes optimal use of the land area, but also improves the efficiency of all subsequent cultivations. Straight rows of evenly spaced seedlings can be cultivated quickly, without the constant stopping and adjusting caused by out-of-place planting. The only way to ensure accurate spacing is to measure. Stretching a

(continued on page 138)

Pre-marking the bed speeds up transplanting.

A right-angled trowel is easier on the wrist, making transplanting a smooth, efficient operation.

Making holes for transplanted artichoke plants with a posthole digger.

One pass lengthwise using a 30-inch (75 centimeter) grading rake fitted with marker pegs to define the rows.

Multiple passes sideways to create the grid.

An electricly powered soil-block transplanter in a Dutch greenhouse.

Efficient Plant Layout—Rake Codes

We mark out row spacing and plant spacing for transplants by using our 30-inch (75 centimeter) bed preparation rake with added marker fingers. The marker fingers are 6-inch (15 centimeter) lengths of red PEX tubing. The transplant instructions to our crew specify a "rake code" for each job—for example "6–7," which is used for head lettuce, kale, and multi-block onions. Each numeral refers to the number of empty rake teeth between each marker finger (one rake tooth equals approximately 2 inches [5 centimeters]).

The first numeral of the pair defines the spacing for the rows running the length of the bed. Whether working with one, two, three, or four rows, we center them on the bed. In the case of 6–7, the numeral 6 results in three rows to the bed about 10 inches (25 centimeters) apart. The rake, with markers at that spacing, is run down the length of the bed to score the surface and define the rows. The second numeral, 7, is the number of empty teeth between the marker fingers for establishing the space between the plants in the row. The rake, with markers set at that spacing, is run across the rows progressively down the length of the bed, leaving a planting grid.

Other examples: The rake code 4–4 gives four rows to the bed for closer spaced crops like parsley, 'Salanova' lettuce, and bunches of scallions. The rake code 10–10 spaces crops like broccoli and cauliflower at two rows to the bed with 16-inch (40 centimeter) spacing between plants in the row. Greenhouse tomatoes and cucumbers, which we set out at 24 inches (60 centimeters) apart down the center of a bed, have a rake code of 8–15.

The bed preparation rake with the marker fingers is a handy system for efficient transplant layout.

tape or a knotted string is a perfectly reliable method (unless a strong wind is blowing), but it is also slow and tedious. A marker rake equipped with adjustable teeth for both lengthwise and crosswise marking is faster. A roller with teeth on it to mark all the plant sites in one trip is better yet.

The Studded Roller

For more efficient transplanting, the next idea is to combine the spacing and hole-making operations in one tool. If a marking roller is fitted with studs that are the size of the soil blocks, both jobs can be done at once. In newly tilled ground, this "studded roller" will leave a regular set of cubic holes in the soil.

A few design modifications can make this idea work even better. The marking studs should have slightly tapered sides (10 degrees) to make a more stable hole. The roller should ideally be 11½ inches

TABLE 18.1. Number of Seedlings and Spacing for 50-foot beds

Number of Seedlings	Rake Code*	Distance (inches)	Crops
500	3–3	6 × 6	Spinach
400	4–3	7.5 × 6	Spinach
320	4–4	7.5 × 7.5	Parsley, 'Salanova,' scallions
300	6–3	10.5 × 6	Garlic, leeks
200	6–5	10.5 × 9	Fennel
175	6–6	10.5 × 10.5	Beet blocks
150	6–7	10.5 × 12	Onion blocks, head lettuce, celery, celeriac
72	10–10	16 × 16	Cauliflower, pepper, 'Happy Rich,' broccoli, cabbage, Brussels sprouts
33	8–11	30 × 18	Pepper
25	8–15	30 × 24	Tomato, cucumber, eggplant

* For more information on rake codes, see "Efficient Plant Layout" sidebar, page 137.

(29 centimeters) in diameter (you should be able to get a local metalworking shop to make one for you). That gives it a rolling circumference of 36 inches (90 centimeters). Then, if a number of stud attachment holes are drilled in the roller, plants can be spaced at 6, 12, 18, or 24 inches (15, 30, 45, 60 centimeters) in the row. The roller should be 30 inches (75 centimeters) wide for the growing area in the 42-inch (105 centimeter) strip. After the ground is tilled, a single trip with the 30-inch roller will prepare the entire strip for transplanting. The final step is simply to set the square block in the square hole. When you place the block, lightly firm the soil around it with the tips of your fingers.

The above is an excellent system, one that I have used myself and have seen in operation on a number of European farms. It has just two small drawbacks. First, if the soil dries out between tilling and rolling, the holes will not form well. Second, the soil at the bottom and sides of the hole is compressed and could inhibit easy root penetration. These are minor points, but they do make a difference. One improvement is to replace the blocks with small (2-by-3-inch /5-by-7.5 centimeter) trowel blades. These are attached to the roller at a 15-degree angle toward the direction of travel. The rotation of the roller causes these "shovels" to dig small holes. Since the holes are scooped rather than pressed, there is no soil-compaction problem.

The next step is to improve the efficiency of the system from two trips over the field to one by combining tilling and rolling into one operation. This is done by mounting a roller with blocks as closely as possible behind the tines of the tiller, after removing the back plate. In this way the holes are formed immediately in moist, newly tilled soil. Compaction is avoided because the roller has become the back plate of the tilling unit. The soil, driven against it by the tines, is falling back into place at the same time the blocks are forming the holes. The roller is attached by arms hinged to the sides of the tine cover. Metal blocks welded onto the bottom edges of the

TABLE 18.2. Plant and Row Spacing for 30-inch Wide Beds

Crop	Rows per Bed	Rake Codes
Artichoke	1	8–15
Arugula	12 rows at 2¼"	direct sown
Basil	4	4–4
Bean	1	8–6
Beet	3	6–6
Broccoli	2	10–10
Brussels sprouts	2	10–11
Cabbage	2	10–10
Carrots	12 rows at 2¼"	direct sown
Cauliflower	2	10–10
Celery/celeriac	3	6–7
Celeriac	3	6–7
Chard	3	6–6
Corn	1 (3 plants per hill)	8–11
Cucumber	1	8–15
Eggplant	1	8–15
Fennel	3	6–5
Garlic/shallot	3	6–3
'Happy Rich'	2	10–10
Kale	3	6–7

Crop	Rows per Bed	Rake Codes
Leek	3	6–3
Lettuce, head	3	6–7
Lettuce, 'Salanova'	4	4–4
Melon	1	8–15
Onion	3 (4 plants per block)	6–7
Parsley	4	4–4
Peas	1 double row at 6"	on center of bed
Pepper	1	8–11
Potato	1	8–5
Radicchio	3	6–7
Radish	12 rows at 2¼"	direct sown
Scallion	4	4–4
Spinach	5	3–3
Tomato	1	8–15
Turnip	6 rows at 4½"	direct sown
Winter squash	1 row per bed @ 72" in row (pair of plants)	—
Zucchini	1 row per bed @ 36" in row	—

tine cover raise the roller when the tiller is lifted at the end of the row.

With this one-pass tiller/hole-maker, plus the convenience of modular plants in soil blocks, an inventive vegetable grower could have a very efficient transplant system for all the crops in 1½-inch (4 centimeter) and 2-inch (5 centimeter) blocks. The 3-inch (7.5 centimeter) blocks and the 6-inch (15 centimeter) pots are transplanted into holes dug by hand. A two-handled posthole digger is the best tool for setting out the 6-inch pots. One quick bite of the jaws leaves a hole the perfect size.

CHAPTER NINETEEN

Weeds

Since the best way of weeding
Is to prevent weeds from seeding,
The least procrastination
Of any operation
To prevent the semination
Of noxious vegetation
Is a source of tribulation.
And this, in truth, a fact is
Which gardeners ought to practice,
And tillers should remember,
From April to December.

—*NEW ENGLAND FARMER,* 1829

I'm always eager to see newly germinated crop seedlings starting to poke through the soil. They are a sure sign that the growing season has begun. Tiny weed seedlings are another sure sign—a sign that weed competition is not far behind. I'm always quick to spot them and dispatch them while they're still small.

Weed Control

There are two conventional approaches to weed control—physical control and chemical control. Physical control involves cutting off the weeds (cultivation) or smothering them (mulching and hilling). Chemical control depends on the use of herbicides. I believe that all herbicides (even so-called natural ones) will be proven to be harmful to both the soil and the farmers.

Physical control, principally cultivation, is the weed-control method of choice here. I emphasize not only cultivation, but cultivation done with hand tools. First, let me stress that this is not the same old drudgery that farm children have always shirked. The tools I recommend have been designed specifically for the job and make it quick and efficient work.

A further emphasis in this system involves more than just the design and operation of the tool—it includes the approach taken by the weeder. Weed control is often considered the most onerous of tasks, and the reasons are obvious. The tools as well as the timing are often drawbacks. Too many growers consider hoeing to be a treatment for weeds, and thus they start too late. Hoeing should be understood, however, as a means of prevention. In other words: Don't weed, cultivate.

Cultivation is the shallow stirring of the surface soil in order to cut off small weeds and prevent the appearance of new ones. Weeding takes place after the weeds are already established. Cultivation deals with weeds before they become a problem. Weeding deals with the problem after it has occurred. When weeds are allowed to grow large and coarse, the task becomes much more difficult. But weeds should not be allowed to become so large. They should be dealt with just after they germinate. Small weeds are easy to control, and the work yields the greatest return for the least amount of effort. As well, small weeds have not yet begun to compete with the crop plants. Large weeds are competition for both the crops and the grower.

The Wheel Hoe

A wheel hoe combines a hoe blade with a wheeled frame to support the blade. Wheel hoes were ubiquitous on vegetable farms in the early 20th century. It is the best cultivation tool for the small-scale vegetable grower. There are two common styles. One has a large-diameter wheel (about 24 inches [60 centimeters]), the other a smaller one (about 9 inches [23 centimeters]). Each style also has different means for attaching tools and handlebars to the frame. I have a strong preference for the small-wheeled model. It just works better.

I remember being told years ago that the advantage of the large-diameter wheel hoe was that it could roll easily over obstacles. My reaction was that, if there were obstacles that large lying about the fields, I had more problems than the selection of a wheel hoe. The truth of the matter is that the design of the large-diameter wheel hoe is faulty. Human power is limited and shouldn't be wasted. In a well-designed tool, the force exerted by the operator is transferred directly to the working part. In the case of a wheel hoe, the working part is the soil-engaging tool, not the wheel. The low-wheeled design transfers force much more efficiently than the high-wheeled model.

Chickweed.

Galinsoga.

Lamb's quarters.

Purslane.

Because the force is direct, a much higher percentage of the effort is applied to the cultivating blade. A further disadvantage of the high-wheeled hoe is that a forward force is being used to manipulate a rear-mounted implement, thus causing torsional (twisting) forces to come into play that put even more strain on the operator. In sum, the low version is more accurate (easier to direct), less tiring (no force is wasted), and less cumbersome to use.

In recent years improvements have been made on this reliable tool. The heavy metal wheel of the old models, with its crude bushing, has been replaced by a lightweight rubber wheel with ball bearings. The original cultivating knives have been replaced by a far more efficient oscillating stirrup hoe, which has a hinged action and cuts on both the forward and backward strokes. When combined, these improvements result in the most efficient implement yet designed for extensive garden cultivation.

Oscillating stirrup hoes are available in widths from 6 to 14 inches (15 to 36 centimeters). Wing models can extend the total cultivating width out to 32 inches (80 centimeters). The curved shape of the stirrup hoe blade cuts more shallowly next to the crop plants than in the middle of the row, thus sparing crop plant roots. The open center also allows

Johnny's new U-bar wheel hoe with leaf lifters.

rocks to pass through and even to be lifted up and out of the soil for later retrieval.

Stirrup hoes are double-stemmed; that is to say, they have a vertical support at each end of the blade. Because of this shape they are most effective when the crop is small or has only vertical leaves. To cultivate around and under the leaves of spreading crops, a single-stemmed hoe is necessary. Single-stemmed goosefoot and chevron-shaped hoes are available in widths from 5 to 10 inches (13 to 25 centimeters).

Old-time wheel hoes were equipped with small plow attachments to turn a single furrow. I have never used one for plowing, but they are quite handy for creating furrows for burying the edges of plastic mulch and floating row covers. Wheel hoes can also be equipped with double-bladed plow bodies for creating a furrow, and with cultivating teeth for soil aeration. Fitting a wheel hoe with two wheels allows it to straddle the crop and hoe both sides at once. When two-wheeled models were common years ago they were connected by an upside-down U-shaped frame that allowed crops up to 16 inches (40 centimeters) tall to pass underneath. This allowed the use of paired sweeps or small disk cultivators to hill soil around the base of established plants to smother in-row weeds. A modern version is now available. Many contemporary wheel hoes add two wheels to a standard low chassis, but without the extra clearance they are not able to cultivate crops more than a few inches tall.

USING THE WHEEL HOE

A wheel hoe equipped with an oscillating stirrup hoe blade makes for pleasant work. The operator walks forward at a steady pace while making smooth back-and-forth push-pull motions with the arms. The push-pull takes full advantage of the swinging action of the oscillating blade and keeps its cutting edges free from debris. Accuracy along a row of seedlings is precise. Gauge your aim by focusing on one side of the blade (depending on whether you aim with the right or left eye) as it passes close to the row of

seedlings. Ideally, the work should be done to a depth of only an inch (2.5 centimeters) or less.

A good wheel hoe will be adjustable at the front forks to change the blade's angle of attack, making it steeper for harder soils and shallower for light soils. The handles can be adjusted for the height of the operator. Set the height adjustment so that your hands are at about waist level and your forearms are parallel to the ground.

A further refinement makes the wheel hoe ideal for bed systems. Depending on the model, these tools are equipped with either a swivel joint or extension brackets through which the handles can be set at an angle off to one side. This feature allows you to walk alongside the direction of travel and avoid stepping on the newly cultivated soil. Thus, when crops are grown in a bed system, all foot traffic can be confined to the paths.

This blending of the old and the new in modern wheel hoes results in the most efficient hand-powered cultivating implement available. It permits a high level of cultivation accuracy (to within ½ inch [1.25 centimeters] of rows of newly germinating seedlings) in addition to great speed of operation.

With the precision seeder or the transplant roller described earlier, you can plant straight, evenly spaced rows. With the wheel hoe you can cultivate those rows right up to the seedlings. When used in combination, these systems do away with the major part of the labor previously required for thinning and hand-weeding. The efficiency of these intermediate systems makes this scale of vegetable production just as competitive as large-scale, highly mechanized, and chemical-dependent systems.

The Long-Handled Hoe

The garden hoe that is well established in the public mind traditionally has a wide blade, a blade-to-handle angle of about 90 degrees, and a broadly curved shank between blade and handle. The working edge of the blade is offset from the line of the handle. It is frequently a crude and heavy tool because it was designed for moving soil, digging, chopping, hilling, mixing concrete, and so on. It is this tool that has given hand-hoeing a bad name. A well-designed cultivating hoe, on the other hand, should have a thin, narrow blade, a blade-to-handle angle of 70 degrees, and a slightly curved shank between blade and handle; the working edge of the blade will be in line with the centerline of the handle (a collineal design). This is a light and precise tool designed for a specific purpose—shallow cultivation. It is used as a soil shaver or weed parer rather than as a chopper or digger.

A number of considerations are important in hoe cultivation. It is undesirable to move excess soil, since this can bury seedlings or throw dirt on plants. The hoe blade should therefore be narrow and thin. The work must be accurate so as not to damage the plants. The sharp edge of the blade should, therefore, be intersected by the line of the handle so the hoe can be aimed accurately. The work must also be shallow and not cut crop roots. The blade-to-handle angle must be set precisely. The technique is to skim just below the surface of the soil, not chop, and the action should be fast and efficient, not tiring. The tool should be light, sharp, accurate, and easy to use.

HAND AND BODY POSITION

When you are using a hoe, stand upright in a comfortable, relaxed position. The traditional bent-over position—and resultant sore back—is a consequence of the chopping hoe, not the cultivating hoe. Body position is determined by hand position on the handle. There are four possible hand positions on the hoe handle: (1) both thumbs up, (2) both thumbs down, (3) both thumbs in, or (4) both thumbs out. The last one is uncomfortable. The third is a compromise. The second one, with both thumbs pointing down the handle, is the conventional chopping-hoe position and results in bending the back. The first position is the way a cultivating hoe is held.

With both thumbs pointing up the handle, you can stand comfortably upright. This is not a new

The comfortable upright position when using a collinear hoe.

technique to learn, only a new application. Most people instinctively hold a broom or a leaf rake in the thumbs-up position. If you think of a cultivating hoe as a "weed broom" to be used with a pulling or drawing motion, the hand position comes naturally. Move your hands in unison to draw the hoe and adjust them separately to optimize the working angle of the blade to the ground. These are very similar movements to those your hands and arms make when you're sweeping or raking. With a little practice, the quick, accurate strokes that so effectively deal with weeds will become second nature to you.

A Razor Edge

A dull hoe blade increases the work and lessens the efficiency of cultivating to a far greater degree than you might think. I have seen a number of estimates (depending, obviously, on how dull the hoe is), but I would say that even a moderately dull edge lessens efficiency by 50 percent. Like any edge tool designed for cutting, the working edge of a hoe (hand or wheel) must be sharpened. A number of different sharpeners will do the job, but I prefer a small file. Carry this in your back pocket and use it at regular intervals to catch the edge before it becomes dull. A hoe blade should be filed to a chisel shape so the cutting edge is closest to the soil.

Tool Weight

An average fieldworker using a cultivating hoe will make some 2,000 strokes in the course of an hour's work. A real go-getter might do twice that. In working with a wheel hoe, about 50 push-pull strokes will be made per minute, and the direction of the tool must be reversed at the end of each row. In any such repetitive work, the weight of the tool is an important consideration. If a tool weighs even a few ounces more than necessary, the effect of moving that weight over a day's work results in the unnecessary expenditure of a great deal of energy. A well-designed cultivating hoe should weigh no more than 1½ pounds (680 grams). A modern wheel hoe should weigh no more than 15 pounds (6.8 kilograms).

When I first became involved with intensive greenhouse salad crops grown in narrow rows, I realized I needed a specialized tool to cultivate the inter-row areas. After pondering the problem and trying every hand-cultivating tool I could find, I went to the workshop. I hammered flat 4 inches (10 centimeters) at one end of a piece of No. 9 wire and sharpened one edge of that surface. I then bent the wire around and back on itself and then off at a right angle so I had a thin hoe blade with an 8-inch- (20 centimeter) long wire shank aimed at the center of the blade. I stuck the shank into a wooden file handle, bent it slightly so there would be a 35-degree working angle to the soil, and returned to the greenhouse. This simple homemade tool represented a quantum leap in successful greenhouse cultivation.

The short blade fits between the rows. It is strong enough for the light cultivation I need it to do. I use it with a slight vibrating back-and-forth motion. This is a perfect example of a tool designed for doing a job in the right place, at the right time, with the right touch. Whereas it would not be at home in a hard soil, the friable greenhouse soil is ideal. It would bend out of shape if I used it to chop larger weeds. It has just the right strength and precision for light cultivation when the newly germinated plants are small. The very thin, sharp blade cuts shallowly, just under the surface, and doesn't dump soil on the crop seedlings. Best of all it is pleasant to use. Both long-handled and short-handled versions of this tool, known as a wire weeder, are now available commercially made from heavier stock to be stronger.

But I kept thinking about the light ⅛-inch (0.6 centimeter) wire we had used for our first prototype, wondering if that thinner wire would be strong enough for harder work if not limited to an L shape. So I put a length of very stiff ⅛-inch wire in a vise, bent the ends around toward me to create a hoe shape, and then bent the ends where the pieces met in the middle at a 90-degree angle back toward me. I then drilled a ¼-inch (6.4 millimeter) hole vertically in a ⅜-by-¾-inch (9.5 by 19 millimeter) bolt, inserted the ends, trimmed them, and braised the opening. I welded a ⅜-inch nut on the end of a hoe ferrule so I could screw on whichever size of wire hoe I wished to use. (We eventually made them 4, 6, and 8 inches [10, 15, and 20 centimeters] wide.) If you transplant seedlings immediately to a just-prepared bed, there is no faster, more efficient way to erase all the small, newly germinated weeds that appear a week later than by using this tool. My farm crew calls it "the eraser." Best of all, since the blade is not sharpened, it does not cut the stems of transplants if you accidentally bump into them. Thus you can erase even the smallest weed that is close to the stem by placing the wire hoe blade against the stem and pulling toward you. Another step toward clean cultivation. If you don't have the option of

The collinear hoe.

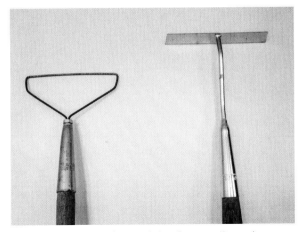

The homemade wire hoe and the classic collinear hoe.

Fitting the Handle to the Collinear or Cultivating Hoe

There is normally some amount of curve in any long wooden handle. For the ideal balance or "feel" of the tool when in use, you must accommodate that curve when attaching the handle to the blade, as follows:

1. Insert the handle into the ferrule.
2. Set the hoe to balance horizontally on a narrow crossbar (for instance, the back of a kitchen chair).
3. When the tool is balanced, note the position of the hoe blade. The desired position is with the hoe blade's cutting edge down.
4. Twist the handle in the ferrule until you find the position where the hoe balances as desired.
5. Set the handle firmly into the ferrule and attach it with a screw. It is helpful to have two holes in the ferrule for the handle-attachment screw. That way a new area of wood is available when the handle dries and has to be pushed into the ferrule slightly farther.

following the description above to make your own version, there is a commercially available model on the market. It is available from Johnny's Selected Seeds and includes an exchangeable assortment of wire blades.

Wide-Row Cultivating

Widely spaced row crops are cultivated most efficiently with the tiller. The choice of row spacing when planting these crops is dictated by the adjustable dimensions of the appropriate tiller model. There should be at least 2 inches (5 centimeters) of clearance between the edges of the tiller and the crop rows. For example, a 26-inch- (65 centimeter) wide tiller carefully steered would just be able to cultivate rows planted at a 30-inch (75 centimeter) spacing. You can set the depth skid for the tiller tines so they run shallowly, ideally no more than 1 inch (2.5 centimeters) deep. For hilling potatoes, adjust the wheels to the narrowest setting and replace the tilling attachment with a furrower.

A recently available attachment for the walking tractor is the power harrow combined with a mesh roller. The harrow teeth work vertically in the soil rather than horizontally like tiller tines. The depth of penetration can be controlled by adjusting the roller. The rotary harrow won't do primary soil preparation as well as a tiller, but it is an ideal tool for early-season cultivating of widely spaced field crops and also for resurfacing newly harvested beds to create an initial weed-free surface.

Flame Weeding

The use of flame or heat to kill weeds in row crops may seem modern, but it is not a new idea. The earliest recorded demonstration of a "weed scorcher" took place at the Royal Agricultural Show in England in 1839. The implement consisted of a horse-drawn, wheeled metal grate (on which a coke fire was built) equipped with a fan (powered by a geared handle turned by a farm laborer), which blew the heat down to kill weeds below.

Flame weeders using liquid fuel were patented in the United States as early as 1852. From then until 1940 numerous models were designed and improved, principally for use in cotton and sugarcane fields. All the early models used liquid fuel, a compressor, a fuel pump, and special generating burners. In the early 1940s liquid propane (LP) gas became the fuel of choice, since it burns cleanly and leaves no residue. From a modest beginning with 10 LP gas flame cultivators in the early 1940s, the concept steadily gained in popularity until, by 1963, an estimated 15,000 row crop machines were in use.

An experimental flamer design. The propane tank rides on the roller.

The practical and economic effectiveness of modern flame weeding depends on a number of factors. The weeds should be less than 2 inches (5 centimeters) tall. For larger weeds, the flaming exposure time can be extended by slowing the tractor speed, but obviously this uses more fuel. Actually, a flame weeder does not burn the weeds. The heat intensity (fuel use) and time of exposure (travel speed) are adjusted so that only enough heat is applied to the weeds to cause expansion of the liquid in the plant cells and the consequent rupture of the cell walls. That requires a temperature of about 160°F (71°C) for a duration of one second. A simple test for sufficient heating is to lightly press a leaf between thumb and forefinger. If the leaf surface shows a dark green fingerprint, the job is done. Following flaming, the weeds continue to look normal for several hours before slowly beginning to wilt and fall over.

Flaming works best on a relatively smooth, clod-free soil, since ridges or clods can deflect the flame up against the crop plants. Flaming is more fuel-efficient when the weed leaves are dry, but it can also be done when the soil is too wet to cultivate mechanically. With the first in-row weeders, flaming was delayed until the crop stem was tall or thick enough to resist accidental flame contact. Optimal speed was about 6 miles per hour. The next development was the addition of water spray shields to in-row flamers to confine the flame heat to the weeds at the base of the plant. The water shields allowed an increased cultivation speed, to 8 miles per hour. Some models combined in-row and between-row flaming, but eventually a combination of in-row flaming and between-row cultivation was found to be the most effective and economical method.

Starting in 1963 the new developments in flame weeding were reported at an annual symposium on "Thermal Agriculture" sponsored by the National LP Gas Association. These meetings continued until 1973, as the fuel crisis that year marked the end of the symposia and of most of the research projects in flame weeding. Fortunately, these ideas were being investigated independently by organic growers in Europe, and they continued to work at it. Higher fuel prices inspired them to make their units more efficient. I first saw a flame weeder on a 1974 visit to organic farms in Europe. Although there were some models developed for in-row use, most of the European growers were concentrating on models for pre-emergence flaming in crops like carrots.

The technique goes like this. After preparing the seedbed the grower waits 10 days to two weeks (during which time weed seeds germinate) and then drills the crop seeds. Just prior to the expected time of crop emergence the field is flamed. Timing is determined by placing a few small panes of glass over the seeded area in selected spots. These provide slightly warmer conditions, so the crop seedlings emerge a day or two earlier. The instant the first seedlings appear under the glass, the panes are removed and the field is flamed to kill the young weeds. The not-yet-emergent crop seedlings are insulated from the heat by the soil. A few days later they emerge through a mulch of dead weeds into a soil that has not been cultivated to bring up new weed seeds.

There are two styles of modern flame weeders— those with uncovered burners and those with burners placed under a metal cover. The cover ideally matches the width of the growing strip or bed in the production system and helps to concentrate the heat. The covered style of burner is more fuel-efficient, using 20 to 25 percent less fuel on average than uncovered burners. Flamers are available in backpack models with a single burner, hand-pushed multiple-burner models (both covered and uncovered) that roll on wheels, and tractor-mounted units. Many growers who are handy at fabricating tools have made their own flamers, since the parts are mostly off-the-shelf items. That fabrication needs to be done with care and knowledge, however, because propane is a very flammable substance. The most important design concept is to balance the number of burners with the size of the tank. The transformation of liquid propane into gas has a cooling effect. If gas use is excessive, the cooling becomes freezing and the regulator can freeze up. Some flame weeder manufacturers try to solve that problem by delivering the propane to the nozzle as a liquid, but that causes other instabilities. My recommendation to growers who wish to use this technology is to check out a flamer from a competent manufacturer. Once you become familiar with its intricacies, you can decide if you wish to fabricate your own.

One experimental flamer design has the burners attached to the front of a 30-inch- (75 centimeter) wide, mesh-surfaced garden roller. The propane tank rides on the weight platform over the roller and the flame is controlled from the handlebars. Five nozzles flame a 30-inch-wide area directly in front of the roller as it is pushed down the bed.

Another flamer design, sold by Johnny's Selected Seeds, consists of a lightweight protective hood supported by a roller. It clamps to the head of a standard backpack flame wand. The hood keeps the heat in and protects nearby crops. At 15 inches (38 centimeters) wide this is an efficient tool for flaming 30-inch-wide beds, whether in the greenhouse or the field, with one pass down and one back.

I have seen many different homemade designs on small farms in Europe. Some of the small one-row models, especially those that roll on a wheel and have crop-protecting guards, seem particularly well suited to meeting specific weed-control needs in specialty crops. The future of flame-weeding technology is just waiting for the ingenuity of practical growers to modify and refine it in ways as yet unimagined.

Another way to kill weeds with heat is to cover the soil with a sheet of clear plastic during the summer in the hope of trapping sufficient solar gain to raise the soil temperature high enough to kill weed seeds in the soil. The technique is called soil solarization. During the summer of 2014 we had an unused greenhouse and decided to solarize the soil in there.

Soil solarization has been investigated since the 1960s in warm-climate areas with lots of sun. We fig-

ured the second layer provided by the greenhouse would make up for our not being in Texas. We followed the recommendations for soil solarization by first irrigating the soil (moist soil transfers heat better than dry soil) and then laying down clear plastic on the soil from edge to edge. Once that was done we shut up the greenhouse for three weeks starting in mid-July. We used soil thermometers to make sure we hit 145°F (63°C) at the 2-inch (5 centimeter) depth to kill the weed seeds. The result was the most weed-free greenhouse I have ever experienced; just amazing, actually. If you read the scientific literature on soil solarizing, the studies show the elimination of many soil diseases as well as a beneficial effect on soil fertility. The soil microorganisms and the earthworms apparently flee downward and then return. The literature doesn't mention any negatives, and we have not noticed any.

Our other large greenhouses, all of which are movable, grew summer crops that year, but we are thinking now that these houses could be moved off those crops in mid-July (we would need to grow non-staking tomatoes) to solarize the next-door plot that will eventually be in winter crops.

We would like to use solarization in July on the outdoor fields also—especially those that we will be planting to fall crops in August. We are experimenting with using two layers of plastic with a slight separation between them to duplicate the greenhouse effect. Possibly we could use solarization as a no-till technique by mowing a green manure crop very short with a flail mower so it would be in a mechanically stressed state before we lay down the plastic to stress it further with heat. We enthusiastically encourage other northern-tier growers to join us in experimenting with soil solarization.

Similar to solarization as a non-soil-disturbance method is occultation—covering the soil with an opaque tarp for a period of time to block sunlight, and destroy weeds by inhibiting photosynthesis. Occultation is slower but may be a more useful technique in cool, cloudy outdoor conditions where solarization is less effective.

Plastic soil cover for solarizing the soil in a greenhouse.

Long-Term Benefits

The encouragement of undersowing in the green-manure chapter obviously limits the period during which cultivation can be practiced. That is why cultivation must be done so well prior to the undersowing date. After that point the expanding leaf canopy of the crop progressively inhibits weed growth. When the additional low-leaf canopy of the undersowing is added, the inhibitory effect is even more pronounced. The few weeds that do manage to grow, however, must be pulled. The grower should plan occasional forays through the garden for this purpose.

That exhortation may seem perfectionist, but with good reason. I remember reading an old grower's book, *Ten Acres Enough*, that was published in the 19th century. The author held a weed-free philos-ophy similar to mine and wryly noted that the neighbors thought he was wasting his time. His conclusion, on the other hand, was that the number of new weeds was smaller every year and that his diligence paid off in the long run. My experience is the same. Ditto Peter Henderson. In *Gardening for Profit*, published in 1867, he wrote:

. . . but weeds should never be seen in a garden, whether it be for pleasure or profit; it is shortsighted economy to delay the destruction of weeds until they start to grow. One man will hoe over in one day more ground where the weeds are just breaking through than six will do if they be allowed to grow six or eight inches in height, to say nothing of the injury done to the ground by feeding the weeds instead of the planted crops. Another benefit of this early extirpation of weeds is that, taken in this stage, they, of course, never seed, and in a few years they are almost entirely destroyed, making the clearing a much simpler task each succeeding year.

I don't worry about the weed seeds in the soil at the start. They are a given. However, I do concentrate on preventing their numbers from multiplying. Since the undersown green-manure crop is a "deliberate weed" that benefits the system, it is the only weed I want. It may not be possible to attain perfection, but I will certainly get a lot closer to that ideal if I prevent weeds from seeding.

Weeds research also confirms this observation and details the rate of progress in ridding crop ground of weeds.[1] A study at the National Vegetable Research Station in England focused on the rate of decline of viable weed seeds in the top 9 inches (23 centimeters) of the soil. The rate of loss varied between 22 and 36 percent per year, depending on the amount of soil disturbance. On undisturbed soil, where no new weed seeds were brought up from below and no new seeding was allowed, scarcely any seedlings appeared after the fourth year. Again I concur from my own experience. After three to four years of diligence, your weed problems should be behind you.

System Benefits

One advantage of any system is the stability brought about by the practices involved. In a well-designed system, the practices add to that advantage by complementing one another. For example, the two planting techniques presented in an earlier chapter aid in efficient weed control. The use of a precision seeder gives evenly spaced plants in straight, evenly spaced rows. That obviously adds to efficiency and ease of cultivation. Crops grown in soil blocks and transplanted into newly tilled soil solve a number of weed-control problems. First, the crops have a head start on the weeds, which have yet to germinate. Second, the crops are set at the desired spacing, thus reducing adjustments for in-row cultivation.

A Final Word

Throughout this book I have stressed the need for the grower to develop keen powers of observation. The use of hand tools and well-designed systems is an aid to keeping on top of day-to-day changes. An English naturalist once made a comment in reference to Robert Burns's poem "To a Mouse, on Turning Her Up in Her Nest with the Plough, November 1785": "Wee, sleekit, cow'rin, tim'rous beastie / O what a panic's in thy breastie! . . ." The naturalist said that, in a modern age, the poem would not have been written, because a driver of a tractor would not have noticed the mouse. The old horse farmer's advantages were not limited to mice and poetry. A grower with two feet in contact with the earth will notice more about the soil, the crops, and the general state of affairs than could ever be observed from the seat of a tractor.

Despite our modern motorized prejudices, hand tools and simple techniques, when designed correctly, are preferable for many operations. Good hand tools and techniques do not represent a step backward. They are, together with the other practices stressed in this book, a step forward to a better vegetable farm.

CHAPTER TWENTY

Pests?

It is an undoubted fact that the principal occupation of almost all agricultural research stations today is the search for ways and means to combat plant and animal diseases and pests. This endeavor becomes more costly each year and appears to be a losing battle. May this not be because the scientists are so obsessed and pre-occupied with sickness that they fail to study health? Most of them appear to be asking, "How can we destroy such and such a pest, or cure such and such a disease?" The question displays a negative approach because the answer at best can only be remedial. But a few research workers have begun to post the positive question—"What is health, how can we promote it and so foster natural resistance?" All the indications so far are that the answers to this question are likely to be far more fruitful in their practical results than anything which agricultural science has hitherto achieved.

LADY EVE BALFOUR,
Plant Health, 1949

This chapter celebrates life. It celebrates the coexistence of growing plants, living creatures, and positive thinking. It explains why I have always found pesticides, whether organic or chemical, to be the wrong answer. The systems of the natural world are elegant and logical. The idea of striving to create life-giving foods while simultaneously dousing them with deadly poisons is inelegant and illogical. Too many books focus on the doom and gloom of potential problems and eulogize the negative. I prefer to focus on the promise and practice of elegant solutions and celebrate the positive. Any supposedly insolvable problems are but a prelude to the next celebration.

Plant-Positive: The Other Side of the Tapestry

When faced with an insolvable problem, I stand it on its head. Then I can reconsider it from an inverted view. Very often a valid case can be made for the obverse position. The history of science records numerous cases of once sacred ideas that were shown to be backward—the Ptolemaic concept of the sun revolving around the earth is a well-known example. Two reversals have been in the news recently. Instead of fearing the big, bad wolf, present-day wildlife managers have come to accept the actions of the predator as intrinsic to the balance of the natural world. The forest fire, once something that "only you" could and should prevent, has reemerged as a necessary component of a healthy forest.

In agriculture, most people would agree that the insolvable problem is the use of pesticides. Even with all the evidence about residue dangers, pest resistance, and environmental degradation, how do you get rid of products that are deemed so indispensable to our food supply? Well, let's turn that one on its head. Instead of the *Pesticides are indispensable, and we can't do without them* attitude that dominates the status quo, the reverse would be: *Pesticides are superfluous,*

Flower borders attract beneficial insects.

and intelligent agricultural systems don't need them. That is certainly an appealing concept, but is there any evidence to support it? Hold on to your hat.

Not only has this concept been documented in scientific studies, but there is also ample practical confirmation of it from farmers' experience. For over a century a small underground of farmers and researchers have rejected the idea that plants are defenseless victims and pests are vicious enemies. In their experience well-grown plants are inherently insusceptible to pests. They contend that plants only become susceptible to pest attack when they are stressed by inadequate growing conditions. Thus, they see pests not as enemies of plants, but as helpful indicators of cultural practices that need to be improved. Simply stated, insects and disease are bringing a message that the plant is under stress. That message is incomprehensible as long as we view pests as enemies. In essence, we have been trying to kill the messenger.

The fact that stress might have a detrimental effect on plants is not surprising in light of the similar effect of stress on humans. When we are under stress we, too, become more susceptible to the ills that can befall us. And just as in agriculture, we can either choose chemical aids to mask the symptoms of our stress or we can make changes to correct the cause—changes in our lifestyle or work environment or daily habits. Any reputable stress-reduction program would recommend the latter as the intelligent course of action.

I define this thinking in agriculture as plant-positive, in contrast with the present approach which is pest-negative. It makes sense. Since there are two factors involved, pests and plants, there are two courses of action: to focus on killing the pest, or to focus on strengthening the plant; to treat the symptom or to correct the cause. Since the former appears to be a flawed strategy, we might be wise to try the latter.

One way to visualize this duality is to picture the natural world as represented by that embroidered tapestry I mentioned in the opening chapter. The pesticide enthusiasts are all looking at the backside of the tapestry. From that perspective they see loose ends, stray threads, and confused patterns. Their science isn't bad, it's just that they can't see the logic of nature's woven fabric. From the front side of the tapestry, the role of agricultural pests as de-selectors of substandard plants is clear, just as vertebrate predators like the wolf are known to target those animals stressed by illness, injury, or senescence.

History and Background

The earliest formal scientific expression of this idea first appeared in plant pathology literature during the 19th century as the Predisposition Theory[1]—that is, the host plant must first be *negatively predisposed* by unfavorable conditions before the pest can prevail. Scientific consideration of predisposition was rapidly eclipsed by new scientific discoveries. Nonetheless, there were a few voices contending that nothing had changed. Even though the microbes could now be identified under a microscope and given names, they could still only prevail against a weakened host. Studies and review articles continued to be published throughout the 20th century. A survey of the literature on one small segment of the subject, the influence of potassium on plant health, cites 534 references, and notes that since 1950 the number of new references available has doubled every decade.[2] A study published in 1984 presenting a theory to explain the relationship between plant stress and insect abundance cites more than 300 references and has itself been cited hundreds of times in subsequent studies.[3]

The general tone of these investigations lends support to what has always been a casually stated but inadequately understood tenet of the organic farming movement: "Healthy plants are not bothered by pests." Or to put it more scientifically, *within a balanced ecosystem plants are inherently insusceptible when properly grown and only become subject to insect and* disease problems when they are stressed by unfavorable growing conditions. In other words, the pest-free plant is not the normal plant with something added, but rather the normal plant with nothing taken away.

Much of the published research is expressed in a confident tone. That confidence is significant in light of today's assumption, quite obviously influenced by modern reliance on agricultural chemicals, that gardeners living in the years BP (before pesticides) must have had only pest-riddled produce, because they were at the mercy of continuous onslaughts of pests. It would appear this alternative understanding was more than a theory at that time but, rather, was the reality of their experience in the garden.

Even Louis Pasteur, whose name is so closely identified with the germ theory, wrote with passion about the potential of this alternative approach to forestall diseases. He was intensely interested in the importance of what he called the "terrain"—the environment within which the organism lives. His greatest fascination was not with the causative role of microbes, but rather with the "environmental conditions" that increased the "vigor and resistance" of the host.[4]

Many investigators have come to similar conclusions regarding insects:

Perhaps in the future more reliance will be put on correct cultural conditions than on spraying, and the conditions of the host plant be more closely watched than the presence of the insect parasite.[5]

Possibly the continuing need for the creation of new insecticides to hold in check greater and more destructive ravages in insect pests is aggravated by the gradual but general decline in soil fertility from year to year.[6]

To sum up, the results already obtained seem to show that the search for improvement in the plant's resistance through its physiology is not just Utopian but quite practical.[7]

How Does It Work?

The most common explanation of how this works focuses on stress-initiated changes in the composition of plants. These changes increase plant susceptibility to insects and diseases. The principal change is a stoppage in the synthesis of protein within stressed plants, which results in a buildup in the plant tissues of free (unattached) nitrogen. Since availability of nitrogenous foods normally limits pest numbers in nature, their populations can explode where stressed plants increase in easily available nitrogenous compounds.[8]

In studies of a rice disease in Brazil, inadequate trace element nutrition of the plant was found to be the key stress. Where specific trace elements were deficient in the soil, the disease was rampant. When the necessary minor elements were supplied, the plants were immune, even under conditions especially favorable to the disease.[9]

Research on nematodes has shown organic manuring of the soil to be important in overcoming nematode damage. The benefits that organic matter confers on the soil—improved structure, greater moisture-holding capacity, increased soil life, better plant nutrition—were all shown to be factors in preventing nematode infestation.[10]

A number of studies point out the difference between "absolute" resistance to pests, which is mostly a factor of the plant's genetic heritage, and "relative" resistance, which is a function of the conditions under which the plant grows. The opinion has been expressed that, if we could learn to nurture the mechanisms of "relative" resistance to their fullest, even practices like large-scale monoculture, often considered a causative factor of insect multiplication, would present fewer problems, because the cultural techniques would have assured the resistance of the plants.[11]

The US Department of Agriculture in its *1957 Yearbook* has this to say: "Well-fed plants are usually less susceptible to soil-borne organisms than are poorly nourished plants. Good fertility may so enhance the resistance of the [host] plant that the parasite cannot successfully attack the roots."[12]

And finally, in one of the few studies specifically relating to organic farming, two Cornell researchers conclude: "Whatever the cause(s) for the significantly fewer insects in the organic treatments, the results support the proposition that organic fertilizers can promote crop-plant resistance to attack by insect pests."[13]

Practical Experience

In the mid-1990s I sent out a questionnaire to 50 of the best commercial organic vegetable growers in the United States. I asked if they had observed a correlation between healthy, unstressed plants and reduced incidence of pests. All but one said yes. I asked them what percentage of the pest problems that affected their chemical-farming neighbors they thought they avoided through growing healthy, unstressed crops. The average response was 75 percent. When you realize these growers have achieved that level of success with no help from the agricultural establishment, the 75 percent figure is quite impressive. It's even more impressive when you realize they are doing the impossible according to conventional agricultural thinking. The majority of these growers agree that the pest problems they still encounter only exist because they haven't yet figured out the successful cultural approach required to resolve those specific stress situations.

I have seen similarly exciting results in growing insusceptible crops on organic farms around the world. One experience stands out in particular. Back in 1979 I was visiting an organic vegetable grower in Germany along with a group of agricultural researchers from the USDA. They were working on a study, *Report and Recommendations on Organic Farming*, which was published by the USDA the following year. One member of the group was an entomologist. While the rest of us held a lively question-and-answer session with the farmer, the entomologist walked into the vegetable field. Stooping over and using his hands to sweep the air above the plants, he surveyed the insect population, looking for pests and pest damage. Eventually the conversation on the side of the field trailed off as our attention focused on the entomologist. He continued his search, becoming ever more amazed at the almost total lack of pest damage to the different crops. Finally, he stood erect, turned to the rest of us, and said in a tone of stunned admiration, "We can't even do this well with pesticides."

Resistance to an Idea

As you can see, there is a considerable amount of evidence—both positive scientific appraisals and successful practical examples—in support of this inverted approach to pest management. It would appear to offer a real option in the pesticide dilemma. Why, then, is this idea totally ignored by mainstream agricultural thought? Could it be that it is so revolutionary in its implications that no one dares to deal with it? Perhaps we are more constrained than we realize by the "mind-forged manacles" of which the poet William Blake wrote. The history of science records numerous instances where erroneous ideas have persisted for decades, even centuries, because of a reluctance to change. In the August 1978 issue of *Ag World*, Lola Smith addressed this very issue:

I cannot understand how [agricultural] scientists must reject so vigorously any suggestion that chemicals may be causing more problems than they are worth as presently used. Perhaps . . . to admit that such is the case would be to admit that the system they have developed with such high hope and optimism may have to be scrapped—and thus a large part of their lives may lose value.

When I suggested to the organizers of two recent conferences on alternative pest control that they

should include at least one paper introducing an outline of this theory, the suggestion was firmly rejected. Their idea of "alternative" was obviously limited to fine-tuning the status quo. I was asking for a revolution. A paper on this subject would automatically cast doubt on the premise of all the conventional papers. If this idea was correct, much of their work would be moot. How could they deal with the fact that I and other organic vegetable growers have been able to establish systems where conventional pest control is not the issue? This revolution deals in realities that are outside the present framework of acceptable entomological thought.

I suggested back in chapter 2 that our scientific language lacks words to describe health. For those wishing to follow this plant-positive thinking in refer-ence to human health, two British medical researchers have proposed the word *ethology*, which they suggest:

> Could be regarded as the study of that state of order and ease forming the background against which disorder and disease become manifest. . . . How lost health can be patched and palliated presents a different challenge to the scientist from how health can be cultivated—grown. . . . These two aspects—pathology and ethology— involve two different scientific adventures.[14]

They go on to suggest that "it is open to us either to promote the Love of Living: or to impose the Fear of Dying." They lament that cure-based procedures are focused on the latter.

Ethology, as currently defined, refers to the study of animal behavior in the natural environment. Perhaps *euology* (from the Greek *eu-*: good, well) might be coined for reference to plants. A new science of Plant Euology would focus on health rather than sickness; on plant enhancement rather than pest control. A Dutch researcher has focused on the phrase *positive health* to stress that health is not a static situation but rather the dynamic ability of the organism to adapt and self-manage. And then there are the words of Aldo Leopold, whose prescient wisdom back in 1938 ties this discussion together. His definition of *health*, "the capacity of a living organism for internal self-renewal," and the British researchers' phrase "mutual synthesis between organism and environment" coalesce with the positive health concept of the Dutch scientist. Unfortunately, as Lady Eve Balfour has pointed out, the science of chemistry, on which most of agricultural science is based, is obviously inadequate to study such a flow of life forces. For the present, both plant health and human health are trapped in a pest-negative world. We can't even begin to seriously consider a concept for which science has no commonly accepted language.

We also encounter a powerfully influenced resistance to this idea based on fear—our fear and mistrust of nature. John Stuart Mill's statement from the 19th century expresses that attitude in unmistakable terms:

> No one, either religious or irreligious, believes that the hurtful agencies of nature, considered as a whole, promote good purposes, in any other way than by inciting human rational creatures to rise up and struggle against them.[15]

That control mentality extends to the words we use. Since insects "attack" the plant and "ravage" the crop, we do "battle" with them in order to "conquer" the "enemy." We use bug "killer" in a spray "gun" to "blast" them. In *The Pesticide Conspiracy*, Robert Van den Bosch paints a compelling verbal picture of the modern pesticide applicator as a swaggering, macho, western gunslinger "pumping the lethal load of his Colt .44" into the bad guy.[16] Our primary view of nature and natural systems is negative. Only rarely do we consider the improved partnership with the natural world that could result from investigating, understanding, and seeking Nature's guidance.

Because we see enemies in nature, even our "alternative" practices are often misapplied. Supplementing the numbers of beneficial insects is one example. As stated earlier, I believe plants are inherently insusceptible when growing optimally in a balanced ecosystem. I try to encourage a balanced ecosystem by planting hedges of selected species and leaving meadow areas interspersed with the vegetable fields, in order to provide habitat for all the beneficial components that help create the balance.[17] If that ecosystem becomes imbalanced, such as by the accidental introduction of a new insect, then searching for and attempting to introduce parasites of that new insect are logical steps toward reestablishing balance. But that practice can be carried too far. Since beneficial insects exist as part of a balanced natural ecosystem, the thought has occurred to use them as a kind of biological SWAT team. Although it may be effective in some cases, the importation and dispersal of predatory insects is based on the same antagonistic thinking as is the use of pesticides. A recent advertisement from an insectary made that quite clear—"Get Revenge with Beneficial Insects." Bringing in mercenary bugs is still pest-negative, still focuses on the "enemy," still treats the symptom, and still attempts to kill the messenger and protect sick plants.

The genetic engineering approach shows less understanding yet. Moving resistant genes from one plant to another is purely defensive and assumes that the natural system is poorly designed. On the contrary, the design is impeccable—it is just poorly understood. A sick plant, even though equipped with a resistant gene, is still a sick plant. When healthy plants are grown under conditions that optimize their well-being, their resistance comes naturally through the proper functioning of all their systems. Genetic manipulation is still a negative rather than a positive solution.

How did we create an agricultural mentality that distrusts and disregards the workings of that very same natural world on which it should be based? In the final analysis it comes down to the fact that we have made nature in our own image. We see natural processes as projections of our own aggressive actions and our revenge-dominated thought patterns. Thus we see malevolence in the relationship of one organism to another and in nature's relationship to us. We don't notice the beneficial balances between predator and prey that are maintained throughout the natural world. We miss the obvious logic of tipping that balance in our favor by creating optimal growing conditions for the plants. We need to shift our thinking.

Making It Work

When I first began trying to create ideal growing conditions, I didn't have to look far for a model to follow. The clues are written on every piece of uncultivated field and forest. The plants growing successfully on a natural site are those whose physiological needs are best met by the soil and climate conditions of that site. Since I wanted the site of my farm to favor a wide range of vegetables, I needed to learn how to make that site as amenable as possible to the needs of each specific crop.

I started with the obvious. I ensured that plant and row spacing were adequate for crop growth, maximum photosynthesis, and sufficient air movement. I avoided planting shade lovers in full sun or moisture lovers in dry soil or acid lovers in alkaline soil and vice versa. Then I dealt with each crop on a step-by-step basis. I would divide a field into strips. Each strip was fertilized differently (say, one with manure, the next autumn leaves, seaweed, compost, et cetera), or each strip received different soil preparation (like rototilling, chisel plowing, mulching,

no-till green manure, and the like). The crops (I usually included more than one variety of each) were planted across the strips.

Where differences were noted, new trials were laid out. The following year saw strips of cow, horse, pig, and chicken manure or beech, maple, oak, and ash leaves or clover, vetch, buckwheat, and rye green manures. Soil that had been deeply aerated was compared with undug; mulch type or depth or time of application was varied. I sent off soil samples for testing to try to pinpoint the beneficial factors involved, and how I might duplicate them by other means. In short, I ran my own experimental farm and developed techniques specific to its conditions. Every year crops grew better, and I had fewer problems.

General experience has shown that practices that stimulate the biological activity of the soil are the most widely effective and least expensive in maintaining the insusceptibility, the yield, and the quality of the produce. These practices include adding organic matter; adjusting the balance, the amount, and the rate of availability of both the major and minor nutrients; correcting the soil pH; aerating the soil to prevent compaction; providing adequate moisture and drainage; using shallow tillage; growing mixed green-manure crops for soil improvement and to modify soil biology; and employing well-designed crop rotations. All things being equal, if I were to suggest just one practice it would be to make as much first-class, well-decomposed compost as possible and incorporate it shallowly. If your efforts are not successful at first, don't give up. You need to adapt your actions to your soil, your climate, and your crops.

In 30 years of growing vegetables, I have never found any need for pesticides once I succeeded in creating the best growing conditions for the crop. The optimal conditions are not the same for all crops, and they are easier to create on some soils than on others. The ideal crop rotation can make a world of difference. So can growing the right variety. I continue to observe and experiment. But in no case

does the creation of those ideal conditions require more than the minimal resources of a small farm, nor more than a reasonable understanding of soil science and agronomic principles. What it does require, however, is a thought pattern that approaches the problem from a plant-positive rather than a pest-negative point of view—from a desire to correct the cause rather than just treat the symptom.

Biological Diplomacy

This cause-correction approach to pests is fundamental to organic agriculture. The hard truth is that if you don't understand this approach, you won't be able to understand how organic agriculture really works; nor will you have any idea of its potential. Without this understanding, organic agriculture continues to be constrained by an imitative type of thinking that merely substitutes "organic" inputs for

chemical inputs. Too many organic farmers unconsciously accept the framework of industrial agriculture, while employing natural ingredients—blood meal for nitrate of soda, bonemeal for superphosphate, rotenone instead of DDT. When done that way, organic farming works reasonably well, because the new ingredients are more harmonious with the natural system than the old ones were. But it hasn't even scratched the surface. A great deal of new research is being published about how "induced plant resistance" to pests and diseases is a result of the intimate partnership that exists between plant roots and microorganisms in a healthy soil.[18]

What I am proposing is a totally revised way of thinking. In order for us to gain a proper understanding of agriculture, we need to develop a biologically oriented thinking that sees our agricultural efforts as participatory rather than as antagonistic vis-à-vis the natural world. This isn't a question of whether pesti-

161

THE NEW ORGANIC GROWER

cides, either natural or artificial, are good or bad. This theory bypasses that unwinnable debate by suggesting that pesticides are superfluous; that they were devised to prop up an agro-industrial framework that was misconceived from the start. When you abandon that framework, you can abandon its negative thinking pattern. The published research and the experience of organic growers around the world demonstrate clearly that when we accentuate the positive, we simultaneously eliminate the negative.

I would be remiss at this point if I did not tentatively extend the discussion one link farther up the food chain to include human beings as consumers of plants. Are we humans also governed by these concepts? As in the case of plants, is our health, vigor, and resistance, our "biological quality," determined by our "growing conditions" and the physiological suitability of our inputs? If we have followed this positive approach to plant health and have optimized all factors of the plant's growing conditions in order to turn out a plant of the highest biological quality, will the consumption of that plant be a factor in optimizing our nutrition and subsequent well-being? The answers to these questions suggest implications of this plant-positive thinking that extend far beyond the field of agriculture.

Almost 500 years ago, while contesting the popularly accepted but flawed Ptolemaic concept of an earth-centered solar system, Galileo realized that he would have "to mold anew the brains of men" in order to establish another understanding. The change I am proposing—from a preoccupation with pest destruction in order to protect sick plants to a focus on plant construction in order to create healthy plants—requires a similar remolding. But it is a change that will allow us to eliminate all pesticides and simultaneously grow more nutritious crops.

CHAPTER TWENTY-ONE

Pests: Temporary Palliatives

In chapter 20, I argued that the emphasis in farming must be redirected toward practices that enhance the vitality of the crops rather than toward methods to destroy the pests. I believe that concept is key to understanding the processes of an ecological agriculture. But as I know quite well, it will be seen only as an "ideal" of pest control by many people who would rather I had concentrated on providing lots of "magic organic solutions" for instant relief. Those in need of such help will find many books specializing in that approach. One quite well-done volume is *The Organic Gardener's Handbook of Natural Insect and Disease Control*, edited by Barbara Ellis and Fern Marshall Bradley (Emmaus, PA: Rodale Press, 1992).

I emphasize preventive thinking because I have no interest in palliatives (from the Latin *pallium*, a cloak). Palliatives are actions that conceal or hide a problem by using a temporary expedient. But I must also consider the opinion of a friend who said to me, "Okay, Eliot, I agree with your plant-positive thinking, but let's get real for a moment. What does a grower do at the start when the systems are not yet together or at those times when things go amiss?" My friend has a valid point. Since very few people have been consciously looking to solve problems from a plant-positive point of view, there is relatively little detailed information available for those facing difficulties with specific crops and specific situations (different soils, climates, seasons of the year). And until such time as attitudes change, individual growers will have little help in a plant-positive quest outside of their own experience.

So for my part, I will suggest below a few pest-negative techniques I have used in those times of need. In my opinion these are the best of a bad lot. I always use them with the caveat that they should be regarded as temporary stopgap measures rather than as long-term solutions. I make no apology for treating "natural" pest-control practices so cavalierly. They don't solve the basic problem. Like chemical techniques, they treat the symptom rather than correcting the cause.

Nutritional Approaches

The nutritional approaches to pest control are actually based on a plant-positive philosophy, and I have no objection to them except for the cost of the materials. In times of plant stress, I have seen benefits from using foliar nutrient sprays to increase the plants' resistance to pests. But the results are not always consistent. In regard to seaweed-based sprays specifically, I think the explanation lies in the cytokinin content of the seaweed product. Cytokinin is a hormone produced by the roots of plants. It has an important function in protein synthesis in the plant. Plants under stress stop producing cytokinins, and thus protein synthesis is inhibited, resulting in insect and disease problems as postulated in chapter 20. Cytokinins applied as a spray to the leaf surface seem to be able to ameliorate this situation. I suspect the inconsistency of results is a function of variation in the time of application (these foliar feeds may be more effective at certain times of day and during certain periods in the plant's development) or the quality of different liquid seaweed products; I cannot say for sure. But I do recommend that growers who plan to use liquid seaweed might want to look for brands that guarantee cytokinin content.[1]

A watery fermentation extract of well-finished compost, used as a foliar spray, has proven effective against a wide range of plant diseases, including potato and tomato blight, cucumber powdery mildew, and botrytis on strawberries. The length of fermentation time appears to determine its effectiveness on different crops. These ideas are developing rapidly, and you should check at the library for the latest research results.[2]

Physical Controls

The floating row covers described in the season extension chapter (chapter 24) work very well as a physical barrier to keep pests away from the crop, providing they are placed over the crop before the insects arrive. We usually support them over pairs of beds with Quick Hoops. Special lightweight weaves designed specifically for pest exclusion, and that have only a minimal temperature-raising effect, are your best bet if the weather is warm. On my dry sandy soil, I have not yet found a plant-positive answer to preventing flea beetle holes in arugula leaves during the summer, and a floating cover has helped a lot.

Transplanting is stressful for plants of the cucurbit family. Plants under stress are more susceptible

Fabric-covered hoops provide physical protection against insects.

to insects. In order to get a longer growing season in our cool climate, we grow our fields of winter squash from transplanted seedlings started in 3-inch (7.5 centimeter) soil blocks. We have found that if the transplants have a few weeks of protection while getting over transplant stress and establishing their root system, they are far better able to shrug off the cucumber beetle. We sow two seeds in each soil block and set out these pairs of plants every 6-feet (1.8 meters) in the row. Since that spacing is not an efficient use of row covers, we developed our own individual screen protectors. We cut 15-inch- (38 centimeter) wide rectangular pieces off a 36-inch- (92 centimeter) wide roll of window screen and fold them into three-sided pyramids with flaps at the bottom that can be covered with soil to hold them in place. We staple the edges and leave the protectors

on for three weeks, until the leaves press against the screen. After removal we store them for reuse in future years.

Vacuum collection of insects is an effective measure against both light-bodied pests (like squash bugs and cucumber beetles) and heavier bugs (like the Colorado potato beetle). Starting in 1979 I conducted pest-control trials using a 5-gallon (19 liter) shop vacuum, and I can recommend it highly for spot treatment of a wide range of insects, especially Japanese beetles.

Back in the 1940s there were two different manufacturers in Texas producing insect-collecting vacuum machines. These machines were attached to the front of a tractor and were driven by a belt running off the PTO. The machines employed the combined effects of both air blast and suction: an air blast from outside

Individual screen protectors.

of the row and a suction intake manifold at the center. Two rows could be cleaned of insects at once, with air blasts from the outside blowing the insects to a centrally mounted vacuum unit. In practice these machines were judged to be as effective as pesticides in controlling cotton insects. Experienced growers suggest that the best time to vacuum is in the early morning, when the insects are sluggish.

Plain water can remedy some problems. Strong, fine sprays of water, especially if they can be directed toward the undersides of the leaves, have been shown to be effective at washing between 70 and 90 percent of aphids and spider mites off plants.

Natural Pesticides

If you read the toxicity data on many natural pesticides, you will learn that it is hard to defend them as safer for humans. Both rotenone and nicotine products are toxic to most animals. Diatomaceous earth contains high levels of free silica, which damages the lungs and can cause silicosis. Precautions should be taken when using any of them. The safety factor with natural products is that they do occur in nature, do not persist, and do not leave human-made chemical residues in the environment. However, that residue-free safety may be compromised by the additives that are used as carriers and sticker spreaders for the natural materials. I recommend considering the residue issue for any products you intend to apply.

I have used rotenone to control potato beetles, and it worked reasonably well. (However, it is no longer available.) The different *Bacillus thuringiensis* (Bt) strains are also effective at the present time against their target insects. A number of new and supposedly safe products have been appearing on the market. I have used finely ground rock powders like basalt dust as an inert pesticide by dusting them over the plants.[3] (Basalt is safer to use than granite dust or diatomaceous earth because it contains almost no free silica, which can damage lungs.) These finely ground powders are very effective in dry weather. In contact with the insect, they either adsorb or wear off the wax layer that covers the insect's exoskeleton, and the insect dries out. There are no residue problems, since basalt dust is also used as a slow-release soil amendment. Unfortunately, most of these materials are not selective and will kill non-target as well as target insects. However, that may be less of a concern if you are treating a onetime problem. A dried, powdered kaolin clay product that adheres to and irritates insects is used similarly.

Pest Resistance

There is a built-in Achilles's heel with any pest-negative product or technique. Its action automatically selects for resistant members of the pest species—those whose unique genetics or behavior make them less susceptible to that particular control. The ability of insect evolution to evade our pest-negative control measures is a far more irresistible force than most people realize. For example, even the most effective insect traps select for those individuals who are not attracted to the trap. Their descendants inherit that unique ability. Our more "technological" practices fare no better. An entomologist friend explained to me recently that the sterile male technique, which had been considered foolproof, is breaking down in the face of the evolution of populations with mating behaviors that exclude the released sterile males, and with the appearance of parthenogenic females. The truth is pretty clear! Pest-negative practices are short-term solutions. The long-term solution involves learning how to grow the plants correctly, so you won't need to resort to palliatives the next time around.

CHAPTER TWENTY-TWO

Harvest

Now that time and effort have been expended to grow first-class, top-quality vegetables, there is one last important step—harvesting. A good harvesting system involves more than just getting the crops out of the field. It must also concentrate on preserving the high quality of the produce until it reaches the customer. And it must do so efficiently, from both a practical and an economic perspective. This is the grower's final exam. All efforts up to this point can be wasted by a careless and slipshod harvesting program.

Well-grown crops make harvesting a pleasure.

Preserving Quality

Vegetable crops continue to respire after they are harvested; that is to say, their life processes proceed as if they were still growing. Unfortunately, since they no longer have roots in contact with the soil to maintain themselves, harvested crops have a limited keeping span. The length of time depends on the individual crop, but the process involved is universal. The higher the temperature, the higher the rate of respiration of the crop and the shorter the keeping time. The grower's aim is to slow respiration in order to maintain all the quality factors—sweetness, flavor, tenderness, texture—that have been achieved by careful attention to cultural conditions during growth. This is best achieved by picking the crops efficiently and cooling them rapidly to slow down the rate of respiration.

There are two parts to the harvesting operation: the efficient organization of the actual harvest and the post-harvest treatment.

Tools and Equipment

Efficiency and economy of motion are important in all phases of the physical work of vegetable growing, but nowhere are they as vital as at harvest. Speed is essential. It keeps quality fresh—and as we have noted, quality determines the market. If the crop is grown well but is not harvested or handled properly, the earlier work was all for naught. Harvesting speed is initially a function of organization beforehand. The

![Hakurei turnips for both roots and greens.]

Hakurei turnips for both roots and greens.

Our washing and packing area is at one end of a greenhouse.

grower must ensure that there are adequate tools—knives, baskets, containers—on hand. The key tool is a good harvest knife. Some growers prefer the California field knife with its large, broad blade. Others use shorter styles with a hook-shaped blade like a linoleum knife or a belt sheath knife with a 3-inch (7.5 centimeter) blade. In many cases the choice of knife depends on the crop to be harvested. For example, I use a field knife for broccoli and cauliflower but a lighter knife for harvesting butterhead lettuce. I also like to have a wrist loop attached to the knife so it remains on my wrist even when I let go of it with my fingers.

Harvest baskets or crates are most efficient if they are of regular size and sturdy enough to be filled in the field and stacked for transport. We presently rely on used bulb crates (purchased from Dutch bulb importers) for all our harvesting. These crates with an open-mesh design are desirable because they are easily cleaned and can be dunked in ice water if desired to quick-cool the produce.

The truck, trailer, or harvest cart for collecting or transporting the produce must be suited to the job. A well-designed harvest cart will have the wheels and support legs spaced so they straddle the growing strips. For this system the wheels would be set on 42-inch (105 centimeter) centers. Heavy-duty cart wheels can be purchased from garden-cart makers or general tool catalogs. The wheel diameter should be 24 inches (60 centimeters) or more. The best "body" for the cart is a flat surface for holding crates. The pickers can then cut and crate produce directly onto the cart as they move down the row. This can be the

Quick-cut. An ingenious small-scale baby-leaf salad harvester.

Unique European sit-down, pedal-powered harvest cart.

Homemade harvest cart made from pipe and simple metal fittings.

Harvest cart with wooden platform is sized to straddle beds.

same cart used for carrying flats of soil blocks during the earlier transplanting operation. It is a good idea to provide temporary shade over the cart until the harvested crops reach the permanent storage area.

On a 1989 trip to Europe I saw some interesting designs for pedal-powered harvest aids. Models had been designed for single workers and for two or three people working in unison. These pedal carts

170

Midwinter spinach harvest from unheated greenhouse.

are supported on fat pneumatic tires, like the tire on a wheelbarrow, which roll in the paths or wheelings between the beds. A formfitting seat, close to the ground, allows the worker to reach efficiently to left or right. Since the seats make for such a comfortable working position, these rigs are an ingenious alternative to the stooping and bending involved in picking low crops like bush peas, strawberries, asparagus, or cucumbers. The units are designed with pedals, chains, and gears for high power and slow speed. They enable the workers to progress along the beds at whatever speed is appropriate to the work being done. Crates and boxes are stacked on the frame. Logically, the units are also valuable for hand-weeding and transplanting. We have purchased one and are experimenting with ways to make it an even more efficient aid to the vegetable grower by powering it with hub wheels and batteries.

Planning

Harvesting involves a great deal of repetitive work. Repetitive work is made much easier and more pleasant when economy of motion is understood and an efficient working rhythm can be maintained. To satisfy those criteria you must evaluate the job from top to bottom. What is going to happen? How is it to be done? What hand and body motions are involved? Is it simpler from left to right or vice versa? It is always possible to find an easier, quicker, and more economical way to do a job. The benefits of such improvements are important in reducing drudgery for the farm crew.

Thus, the keys to simplifying farmwork, especially harvesting, are to:

- Eliminate all unnecessary work.
- Simplify hand and body motions.

171

- Provide a convenient arrangement of work areas and locations for materials.
- Improve on the adequacy, suitability, and use of equipment needed for the work.
- Organize work routines for the full and effective use of labor and machines.
- Involve the workers in the process. When people become more conscious of the way they perform work, their interest increases and their attitude toward the work changes. They begin to notice other things and make valuable suggestions for further improvements.

Minor Details

Let's take tomato picking as an example. Studies have shown that the average worker does not need to work harder, but rather more efficiently. Comparative trials have demonstrated the difference in worker productivity that can result from very simple changes. A comfortable handle on the picking basket so it can be moved with one hand rather than two may seem like a small detail, but the increase in efficiency is considerable. Picking with both hands and keeping them close enough together so the eyes can control them simultaneously without moving the head speeds up the process. The hand motion itself is more efficient if two tomatoes are grasped instead of just one. Since 40 percent of the picker's time is spent moving the hands to the basket, the picking rate can almost be doubled by learning the finger dexterity needed to pick two fruits at once. The technique is to pick a tomato in each hand, shift the tomatoes back into the palm of the hand, and pick a second tomato in each hand before moving the hands to the basket.

The upshot of approaching the physical aspects of harvesting in such a planned and organized manner is not just an increased speed of one particular task such as tomato picking, but improved efficiency of all harvest work. Further, once the grower and the harvest crew become aware of the possibilities for making the harvest easier and more pleasant by focusing on everything from individual motions to overall organization, the improvements carry over into other aspects of the farm day. Any work that can be done in less time and with less effort is more pleasant and relaxing. Any time spent thinking about and reorganizing for work efficiency is time well spent.[1]

A Shining Example

The harvesting of baby-leaf or mesclun salad on small farms was still done by hand with a sharp knife or scissors for many years after the large California farms had mechanical harvesters. However, thanks to the combined efforts of a number of individuals, a handheld baby-leaf harvester, powered by a cordless drill, finally came on the market in 2012 to make life easier and harvest more efficient for local growers. Called the Quick Cut Greens Harvester, it cuts with a reciprocating serrated blade and collects the greens gently with a powered sweep made of knotted macramé cords.[2] This is a wonderful example of the type of small-scale tool ingenuity stimulated by the needs of the growers that has kept hardworking small farmers competitive for many years.

Another solution, part of the delightful balance between biological solutions and mechanical solutions, are the multi-leaf lettuce varieties such as the 'Salanova' types that can be harvested with one quick knife cut, regrow for a subsequent cut, and have a substantial flavor advantage because the leaves are produced from a more mature head of lettuce.

Post-Harvest Treatment

The best and most complete information on post-harvest treatment of all crops is contained in the excellent publication on the subject by the USDA—*The Commercial Storage of Fruits, Vegetables, and Florist and Nursery Stocks* (Agriculture Handbook

Our root cellar.

Onions sun-drying before the move to the curing house.

Our homemade CoolBot cooler.

Number 66). I have followed their guidelines for many years, with great success.

Harvested produce should be neither immature nor overmature, because in either case eating quality and storage life are impaired. Any non-edible portions such as carrot tops or extra cabbage leaves should be removed unless they are absolutely required for dressing up a sales display. Such large, leafy expanses present extra evaporative surface, hastening water loss and loss of overall quality. The leafy vegetables that wilt the fastest, such as lettuce and spinach, should be harvested during the early-morning hours and taken into cool storage immediately.

The first step after harvest is to precool the crops. Precooling refers to the rapid removal of field heat. Crops harvested early in the morning, before the sun warms things up, have less field heat to remove. Since any deterioration in crop quality occurs more rapidly at warm temperatures, the sooner field heat is removed after harvest, the longer produce can be maintained in good condition. Precooling can be done by immersion in or spraying with cold well water, the colder the better. A superior method, although it is more expensive, is to place crushed ice within containers in direct contact with the produce or spread over the top of it. Freezing units to make crushed ice sometimes can be bought used from restaurant suppliers.

After precooling, any produce to be stored should be kept cool in a springhouse, root cellar, or refrigerated cooler. A temporary homemade "refrigerator" cooled by evaporating water can be set up quickly by using a fan to draw air through a water-soaked cloth into the storage area. This will also increase the humidity around the stored crops. High relative humidities of 85 to 95 percent are recommended for most perishable horticultural products in order to retard softening and wilting from moisture loss. Since most fruits and vegetables contain between 80 and 95 percent water by weight, wilting can seriously lower quality. The least expensive electrically powered walk-in cooler option is to purchase a CoolBot controller. The CoolBot will allow you to transform any insulated room into a walk-in cooler by modifying an off-the-shelf air conditioner at much less expense than a commercial cooler.

At times it may be necessary to store different products together. In most cases this is no problem, but with some products there can be an unwanted transfer of odors. Combinations that should be avoided in storage rooms are apples or pears with celery, cabbage, carrots, potatoes, or onions. Celery can also pick up odors from onions. Ethylene damage is another consideration. Lettuce, carrots, and greens are damaged when stored with apples, pears, peaches, plums, cantaloupes, and tomatoes because of the ethylene gas that the fruits give off as they ripen in storage. Even very low concentrations of ethylene may produce adverse effects on other crops.

CHAPTER TWENTY-THREE

Marketing

There are many marketing options for the quality-conscious small grower. The standard possibilities are restaurants, farm stands, and farmers markets. These are all tried-and-true outlets, and there are excellent real-life models available to study. At one time or another, I have used all of them.

Plan to Succeed

To do successful business with restaurants, the grower needs hustle and dependability. Hustle, in order to find potential restaurant customers and convince them to buy from you, and to offer new crops, extended-season production, and gourmet items in order to increase the business once it is developed. Dependability, to keep that business once you have it by never defaulting on a promised order. Good chefs love good ingredients. You should let them know how nice your crops are by hosting an open-field day at your farm and inviting the chef of every local restaurant. Or else pack a gift basket of your choicest items and deliver it to them with a

Contrasting colors make for great presentation at market.

Maine artichokes. Eat your heart out, California!

clear list of what is available and when. If you have exceptional produce, you will not be begging for customers. In fact, you will be doing the chefs a favor by letting them know it is available.

The secrets to success at farm stands and farmers markets are attractive surroundings, ease of access, cleanliness, orderliness, cheerful service, and early produce. Access to the parking lot must be open and inviting. The general level of cleanliness, neatness, and organization will be the first thing a customer notices. Make sure it speaks well of you. The employee or family member on duty at the stand must be friendly and informative. Make sure it is a pleasure for customers to shop there. Say yes to special requests whenever possible. Always make amends for any customer complaint. Our guarantee was ironclad. We gladly offered money back, replacement of the vegetables, even double money back without argument. It is important that customers know you stand behind the quality of what you sell. That policy never cost us more than $10 a year, and it gained us priceless customer goodwill. Nothing is more expensive for a retail business than unhappy customers.

We often planted perishable crops such as lettuce in the fields closest to the stand so customers could choose their lettuces and have them cut fresh on the spot. Free copies of the vegetable recipe of the day were posted to whet our customers' appetites. Customers were encouraged to wander along the harvest paths and view all the crops. Their presence encouraged us to keep up on our cultivation so the place looked orderly and presentable. A prominent herb garden inspired gourmet cooks.

We wanted our operation to stand out from all the others. We established a reputation for having everything all the time. We pursued that policy by raising the broadest possible range of crops and by using succession plantings. The broadest range of crops meant some 40 different vegetables. Succession plantings meant planting as often as necessary to ensure a continuous supply of each crop from the time it first matured till the end of the growing season. One year, by dint of diligent succession planting and careful mulching, we succeeded in selling fresh peas every day but one, from when they first matured in June till we closed in October. I won't say we made money on the peas. In fact, they were a loss leader. But just having them every day enhanced our reputation and our business.

All of these efforts paid off. New customers would often tell us they had been assured by friends that if anyone would have this or that crop (whatever they were looking for), we would. No other local growers bothered with minor crops like radicchio, scorzonera, or fennel. We did, and we gained a lot of customers by doing so. In addition, all our fields were very neat, trim, and well cultivated. Customers loved the look of the place and would come twice as often and bring friends just because of that. A typical comment was, "Our friends the Smiths came to visit us, and we told them one of the first things they had to see was your farm. You make vegetables look so beautiful."

Another marketing approach is to specialize in high-demand crops such as mixed salads or in a specialty such as winter-season production—two areas we are involved in at the moment. I think the new awareness of vegetables and salads as integral parts of a healthy diet is a movement that will continue to grow. Local salad production can be especially lucrative with an extended season of availability. Once I get customers interested, I want to be able to keep supplying them. More than anything else, local salad vegetables bespeak freshness, crispness, and purity. What a successful drawing card that is for any small-farm operation.

Whatever the style of marketing, presentation is crucial. Potential customers will quickly become aware of the quality of produce you offer if your high production standards are matched by the inviting way you present your produce. The following paragraph from a 1909 book, *French Market Gardening* by

John Weathers, shows the timelessness of this good agricultural advice:

> Perhaps one of the most difficult problems connected with commercial gardening is the disposal of the produce at such a price as to yield reasonable profits. In this connection much depends not only upon the way the "stuff" is grown, but also upon the way it is prepared for sale. It is well-known that the very finest produce in the world stands a very poor chance of selling at all, unless it is packed in a neat, clean, and attractive way . . . Originality, combined with neatness and good produce, very often means remarkably quick sales.

Farmers Markets

When we started going to farmers markets, we knew we wanted a quick, efficient way to set up and break down. Our "veggie wagon" was conceived as a way to do just that. It has been a great success. By being self-contained it has also allowed us to occasionally start instant markets in underserved areas. It was initially constructed from drawings scribbled on odd pieces of paper, as first-time ideas usually are.

We started with a 5-by-8-foot (1.5 by 2.4 meter) flat-bed trailer. We used standard 2×4s to construct the frame of the building. The boarding is tongue and groove for stability. We bolted the structure firmly to

(continued on page 180)

Our mobile veggie-wagon farm stand.

Shelves fold out on both sides for great display space.

We can start farmers markets in any underserved area.

All folded up for transport.

Flowers on a Small Vegetable Farm

By Barbara Damrosch

About 10 years ago flowers started to creep in at Four Season Farm. As our name implies we grow and sell produce in every month of the year, even though our main focus is supplying year-round Mainers with fresh winter food. A flower crop dovetails perfectly with that plan. In July and August our area swarms with vacationers and summer residents. And since we must maintain a large enough summer crew to grow crops for winter storage, flower sales help pay those salaries.

Flowers make a good addition to any small farm. They increase plant diversity and attract pollinators. They give an extra measure of beauty to a farm stand, a market booth, and the farm itself. Eliot and I think vegetables are gorgeous, too, and can more than pay their own way. But why pass up a very profitable crop that people love, and that's such a pleasure to grow?

Introducing flowers was easy, because a whole farm infrastructure was already in place: fields, greenhouses, tractors, irrigation equipment, delivery van, propagation greenhouse, and part of a root cellar

available for dahlia storage. We even have an "onion room" in our barn that can be cooled down in fall for onion and squash storage, thanks to a CoolBot setup. In summer it's used to chill flowers after picking.

Expenses for flowers are few: seed packets each spring, sometimes new tulip bulbs in fall, and several largely reusable items such as Hortonova netting, support stakes, and plastic flower cups. Perennial plants we buy repay their cost, since they multiply year after year. Inputs such as mulch hay are bought at bulk price for the whole farm.

There are many ways to have a flower business. It can be small or large, wholesale or retail—or both. You might start a flower CSA, or decorate restaurants and stores. You could do weddings or just supply brides with bulk flowers to arrange on their own. You might grow dried flowers and turn them into wreaths for winter décor. It depends on what best suits your own particular talents, time, and resources. It also hinges on what kind of product your area will support.

Staked dahlias and zinnias in the flower greenhouse.

One-of-a-kind bouquets, 14 to 16 inches (36 to 40 centimeters) tall, have worked best for us. I enjoy making them, and they sell very well at our on-site farm stand, at farmers markets, and at several businesses where we wholesale them. When selling retail we put them right next to the point of sale to encourage a last-minute purchase. We supply businesses with display boxes that show them off.

Sales begin in mid-March. Flowers are slow to bloom in midcoast Maine, where spring is cold and summer cool. I'm lucky to have the heated propagation house to grow transplants and pot up dahlias for a head start. In addition to outdoor growing areas, I have part of a small glasshouse for early-spring bulbs, and a 50-foot (15 meter), unheated, movable flower greenhouse that provides extra weeks of bloom in spring and fall. Soil blocks give me super-healthy starts with no transplant shock. (In mid-November the flower house slides over a winter crop such as spinach or leeks.)

Having arrived at a price that gives us a profit but still keeps sales going quickly—to keep the flowers fresh—the current goal is to continually streamline the business, growing better and better flowers, and more efficiently. Every year we come up with a few new tricks to make the work go faster. For instance, I'd been growing the flowers the same way we grow our other crops, in 30-inch- (75 centimeter) wide beds with 12-inch (30 centimeter) paths between them. But many flowers are too bushy for that, and were too crowded for easy picking, so I upped some of the beds to 36 inches (91 centimeters), and all of the paths to 24.

Every year I evaluate each flower's performance and the quantity I should grow, trying not to waste space and maintenance time. I'm always looking for better varieties, with great colors, improved vase life, and longer, stronger stems. Though dahlias and zinnias are my mainstay, there are hundreds of flowers I could use to fill a bouquet, so I try some new ones each year. Landscape plants around the house, such as mountain laurel, have proved very useful, as have wild plants such as goldenrod, growing in the non-cultivated areas of the farm. My current favorite is a wild shrub called hardhack, whose fuzzy white or pink spires are the perfect vertical accent in summer bouquets. It is considered a nuisance by most farmers, who hack it out with great fervor. Who knew that it could be a prize crop?

Staked zinnias in one of the outdoor production areas.

Late-June bouquets.

the trailer with metal strapping at all four corners. We added 2×4s in the metal slots at the front of the trailer for extra strength in case of a sudden stop. To keep it light the roof is not boarded but rather just covered with a single layer of overlapping pine clapboards. The roof and trim are painted. Originally the sides were varnished, but that didn't weather well, and they have since been painted. The vegetable display shelves are hinged at the bottom and supported by lengths of chain when open. The sales table folds tight against the back of the structure when it is all closed up for transport.

A removable mini-door covers the bottom of the door area at the rear and is used as a sales shelf for extra flowers sitting on the trailer tongue at the front of the structure. The awning bars and fixtures are made from ¾-inch (2 centimeter) electrical metal tubing (EMT). The diagonal awning supports are removed and stored inside for transit. The awnings are rolled up on their bars and tied to dangling cords attached to the structure. The green-striped awning over the sales table folds down flat over the rear when the sales table is folded up.

A wire brings power from the towing vehicle's cigarette lighter so we can plug in the cash register. A FOUR SEASON FARM sign runs along the peak of the roof so we can be spotted from a distance. The building can carry about $3,000 of assorted fresh vegetables in stackable plastic bulb crates. On arrival we fold down the sales table, set the awnings, and open out the shelves. Then the crates of produce are slid on to the shelves. Extra crates inside replace those that sell out. At the end of market, everything fits back inside quickly, and we head home.

Subscription Marketing

This is a very innovative approach. It has the potential to be quite economically successful for the small

180

Our farm stand.

farm, more satisfying to customers' needs, and less costly in terms of farm labor. I initially called my own subscription program many years ago a "Food Guild." I chose the word *guild* because it is defined as "a voluntary association for mutual benefit and the promotion of common interests." That is an accurate description of this idea. This marketing system is a farmer-consumer symbiosis, a relationship that benefits both parties.

I first heard of this concept in 1980 from a member of the USDA Organic Farming Study Team, who had recently returned from an information-gathering tour of organic farms in Japan. It seems that many Japanese organic producers, who farm on the scale that we are talking about here, found that the best market was a limited group of loyal customers. The farm unit becomes the complete food supplier to whatever number of families the farm production can accommodate—not just for a few products, but to supply all the vegetables and any other farm foods over the course of the year. The customer families, who sign up in advance, were encouraged to become involved with the farm. Everything from choosing varieties of lettuce to determining the number of roasting chickens per year was done in consultation with the customers. Thus they became aware of the source of their daily food, to the degree that many would voluntarily show up in their spare time to help out on "their" farm. The farmer benefits because marketing is no longer a time-consuming process.

Obviously, if one requires this sort of devotion from consumers, they have to receive an equal value in return. And they do. The Japanese use the poetic phrase, *Food with the farmer's face on it*. I have heard other people state that the only way to be sure of eating pure food is to know the first name of the grower. Local organic producers offer customers safe food in an increasingly chemicalized world. Customers will

181

sign up because they can be sure that the farm's food is pure and grown with meticulous attention to detail. They can relax and enjoy eating, free from concern about problems of chemical residues.

A few years ago I met with a German grower who operated a subscription program for 600 households: *From our garden to your house*. He spoke about what it meant to him as a farmer to market in this manner. There was the obvious advantage of knowing the crops were sold ahead of time so he could concentrate on the part he loved best—growing exceptional produce. But even more important than that was the implicit recognition by the customers of his skill at growing. By signing up with his farm, consumers were choosing his services on the basis of their assessment of his professional competence, in the same way they selected a family doctor, lawyer, or other purchased professional service.

PRODUCER-CONSUMER CO-PARTNERSHIPS

The Japanese name for this marketing concept is *teikei*, producer-consumer co-partnerships. The British call it a box scheme. I have learned of many examples of this style of marketing in European countries, where the idea has a long history. The first farm employing this concept in the United States was a product of the European influence. Its founders coined the name *community-supported agriculture* (CSA for short), a name now often applied to all marketing schemes along these lines. Although in most cases the basic concept is the same—the farm has contracted with a group of customers to provide

Attractive display increases sales.

Our favorite way to sell brussels sprouts.

them with a broad-based diet of as many farm foods (usually vegetables, but sometimes also eggs, milk, and poultry) over the course of the year as the farm wishes to produce—the ways in which that idea is presented and marketed can be very different.

The European model that inspired the first CSA in the United States arises from a background of social and philosophical concerns about food, food systems, participation, and human responsibility. People who share those concerns are logically attracted to that model. But there is a huge pool of potential customers whose participation may be inhibited by a focus that goes beyond food shopping. All they want is dependable access to a supply of fresh, wholesome food. Since the first rule of marketing is to give customers what they want, I suggest

that there are many diverse ways in which this concept can be presented and marketed in order to expand its potential.

WHAT'S IN A NAME?

For marketing success and to better capture the essence of what they are selling, different plans may want to use a more precise name than CSA. My original term, Food Guild, is one example. Or it might be better to key the name to a familiar concept. Since people commonly start or join clubs for group enjoyment of a limited resource, the "Organic Food Club" might catch their interest. People with money to invest but no skills in the investment world often turn to a mutual fund. Maybe people with a hunger for local organic vegetables but no skills in

Customers get to see beautiful crops.

farming might be enticed by a "Mutual Farm." Who wouldn't love to have their own private gardener like the lord of the castle? Well, an operation called "The Estate Garden" could advertise its service and produce quality as the equal of having a private gardener, yet be no more expensive than the food store since the cost is spread among many members. Possibly just a local town or region name may be all that is needed: "Mountainside Home-Grown Vegetables, Inc.—sign up now and reserve your share."

The variations, refinements, and possibilities of a guild marketing program are unlimited. There are some approaches that may initially seem more workable than others, but the real determinant will be the desires of the guild participants. Some areas of the country and some groups will require entirely different arrangements than others. What we have here is

a system of marketing in harmony with the biological diversity of farming itself. There are as many marketing choices as there are agricultural choices to accommodate different soils, climates, and locations. It is refreshing to think that this program could potentially be as individual to the farm as are the production practices themselves.

THE POSSIBILITIES

A program can be set up for any degree of participation or non-participation by the customers, depending on the desires of a particular group.

- The customers could help with harvesting and distribution as in a food co-op, or the farm could hire interested customers or outsiders for picking and packing.

Crops displayed inside the farm stand.

- The vegetables can be made available "as picked," or the farm can wash and bunch for a more professional presentation.

- The customers could come to the farm (say, twice a week) to pick up their food supplies, or the farm could deliver to a centralized pickup spot (in a town or city) or to individual customers on whatever schedule was selected.

- The farm could provide just the raw materials, and the customers would be responsible for any processing. Or the farm could freeze or can the storage items for the customers and provide them with the finished products.

- The customers could store out-of-season foods in their own freezers and cellars, or the farm could provide bulk facilities for freezing and storage that the customers could then draw on as needed.

- There could be a specified list of products supplied each week throughout the year, and customers could have the flexibility to request greater quantities of one or the other item, either by paying more or by trading off against something else: more Chinese cabbage, less lettuce; more chicken, less pork.

- The length of the fresh produce season can vary. Growers in a summer-vacation-home area may find a perfect match between the outdoor production season and when potential customers are in residence. If customers are available, the production season can be doubled with simple greenhouse protection at either end. Obviously, the longer the season of availability, the more attractive the program is to potential customers.

- A program could also be conducted in partnership with another organization. Back in the late 1970s, we set up such an arrangement with a nearby food coop. We called it the "organic grab-bag." It was a bag of vegetables at a fixed price, but with no guarantee of the contents other than that it represented a great value for the money. People would sign up in advance, and we would fill the bags with whatever kinds of crops were extra plentiful that week. We deducted an additional 10 percent when the co-op began supplying the pickers (who received co-op work credit for their participation). It was a wonderful arrangement. We got paid for excess crops that might otherwise have gone unsold. The buyers got a great value and took care of the harvesting, packaging, and delivery. All we had to do was show them where to pick and provide a minimal amount of instruction. The opportunity exists to make similar arrangements with any organized group.

- From my marketing experience I would suggest that the more services a subscription program provides, the more attractive the program will be in a world ever more attuned to supermarket convenience. Although as a producer I know that quality is my first consideration, I realize that for many potential customers—those who are not yet aware of differences in quality—service and convenience rank higher. If I want their business, I must take account of that reality.

Pricing

The choice of the either-ors above will determine the price of the food and the level of service. Logically, the more services the farm provides, the more the food will cost. With the exception of the most personalized service, the cost should be similar to standard store prices for comparable items. And that is another benefit of this concept. Organic foods are often criticized as being too expensive for most budgets. That is often a function of the laws of supply and demand in the marketplace. But because of the benefits that accrue to the farm and farmer through this prearranged system of marketing, organized consumer groups can obtain the best organic produce at standard food-store prices.

However, in those cases where items are more expensive to produce, their sale price must cover the increased costs. Since this is a mutually beneficial relationship, the customers will be cognizant of the vital importance of a local farmer to their own happiness and well-being. A price must be agreed on that will allow the farmer and farmworkers a realistic income for providing such an important ingredient in the lives of their customers.

The biological production technologies I recommend are designed to keep small producers in business by lowering their costs. The resultant higher quality of the crops should further aid small producers by increasing their income. Food prices should reflect quality, just as prices do in other consumable goods. There has to be a premium paid for quality and the skill and caring that creates it or it won't exist. The small farmer who is turning out a premium product must demand a fair return.

Stay Small

These days, when a business succeeds there is always the tendency to multiply the success by getting a lot bigger. I have one word of advice—don't. That admonition may sound heretical given the dictates of modern economics, but my experience confirms it. I have seen too many successful vegetable growers make the overexpansion mistake. Without exception, they have each become just another company trading on the reputation they established before expanding. If demand exceeds supply, bring the two back into line by raising prices. Income will increase just as it would by expansion, but quality will not be compromised.

CHAPTER TWENTY-FOUR

Season Extension

We live in an age of supermarket thinking where customers have come to expect out-of-season produce. The supermarket sells tomatoes in April and peas in October. In order to compete, local growers should attempt to come close to those goals for the period of the year when their marketing operation is open. For almost all vegetable crops, a longer growing season is desirable if it can be attained economically. Vegetable growers can capture and hold new markets and receive higher prices by having produce available as early or as late as possible compared with unprotected outdoor crops. The grower meeting a local demand or running a market stand will find that a policy of "everything all the time" pays off handsomely.

The secret to success in lengthening the season without problems or failures is to find the point at which the extent of climate modification is in balance with the extra amount of time, money, and management skill involved in attaining it. When

We use both low tunnels and hoop houses.

planning for a longer season, remember the farmer's need for a vacation period during the year. The dark days of December and January, being the most difficult months in which to produce crops, are probably worth designating for rest, reorganization, and planning for the new season to come. In any season extension, the aim is always to keep the systems as simple and economical as they can be without relinquishing the dependable control necessary to ensure the success of the protected cultivation. A broad range of options is available. This review will run through most of them and then recommend those that fit best into a small-scale vegetable operation.

Climate Modification

When considering the possibilities for extending the growing season, we should be aware of options other than building a greenhouse or moving south. There exist many low-cost or no-cost practices that can make a significant contribution to modifying climate conditions in the grower's favor. Any human modification of climate involves altering the existing natural parameters in order to achieve more than those parameters would otherwise allow. As a rule of thumb, the more we wish to modify the climate, the more energy we must expend in doing so. For example, growing tomatoes in January in New England obviously requires a far more expensive and extensive effort than merely speeding up the ripening of summer tomatoes by setting out the plants in a particularly warm and sheltered spot.

The simplest way to improve the growing conditions for early crops is to find, create, or improve such warm and sheltered spots.[1] We can temper the climate with minimum energy expenditure if we work with natural tendencies and try to augment their effects. Three common natural parameters

Cucumbers on one side, tomatoes on the other.

Newly established cucumbers.

that are logical candidates for modification are the degree and direction of slope of the land; the amount of wind exposure or protection; and the heat-absorbing potential of the soil (soil color). If your land does not slope to the south, a practical approximation of a southern exposure can be created on a flat field by hilling it up in east-west ridges with the southern surface sloped at approximately 40 degrees. This gives the effect of many small south-facing slopes. Calculations have shown an average 30 percent gain in total heat absorption by the soil from this practice. Early crops planted just up from the base of the southern slope of these ridges have a significant head start in the spring over similar crops planted on the flat.

Windbreaks of hedges and trees require long-term planning if such shelter does not already exist. Fortunately, effective short-term protection can be had by using temporary shelters. Two parallel lengths of snow or dune fencing 4 feet (1.2 meters) high and 60 feet (18 meters) apart can increase air temperature by 1 to 4°F (roughly 1–2°C). The slightly more substantial protection of a snow or dune fence combined with a hedge can mean an average air temperature increase of 5°F (3°C) in April and 7°F (4°C) in May. Even a temporary 2-foot (60 centimeter) hedge of spruce boughs can keep the air temperature of the protected plot 2 to 3°F (1–2°C) higher than that of nearby open fields.

On small European farms I have seen temporary windbreaks of woven mesh materials placed on the edges of fields perpendicular to the prevailing wind and also within fields to create smaller protected areas. A number of companies sell these specialty windbreak fabrics, which have securing grommets already set into the fabric edges. They are usually made in 4- or 6-foot (1.2 or 1.8 meter) widths and come in lengths of 100 to 300 or more feet (30 to 90

Greenhouse crops are grown vertically for best use of space.

meters). They are attached by the grommets to wooden posts set at a 6- to 10-foot (1.8 or 3 meter) spacing depending on the wind conditions. In really windy conditions the fabric should be secured to the post with a full-length batten. These windbreaks provide part of the advantage of a greenhouse. You have the walls and not the roof, which means that irrigation and ventilation are still provided by rain and natural air movement. Yet the temperature around the plants has been raised substantially. The closer together the vertical walls are placed, the more protection they give. Temporary 2- to 3-foot- (60 to 90 centimeter) high walls of translucent plastic between tomato rows, for example, will raise the daytime temperature 10°F (5–6°C) on average, with no worry about overheating as in a covered greenhouse.

Soil color is a third factor that adds to the effects of the first two. The beneficial out-of-season growing conditions created by the combination of a southern slope and wind shelter can be augmented even more by darkening the soil. Certain soils warm more rapidly than others, and this natural power of absorption can be increased. Dark colors absorb more heat than light colors, and soils are no exception. Charcoal, carbon black, and coal dust have all been used successfully to darken and increase the heat absorption of soils.

In an experiment conducted to test the way in which soil color affects the rate of heat absorption, three plots were prepared on a natural sandy loam. The first was covered with soot to make it black, the second was left with its normal soil color, and the third was dusted with lime to whiten it. Thermometers were placed at a 4-inch (10 centimeter) depth in each plot. By midafternoon on a sunny day in early May, the blackened surface had raised the temperature at the 4-inch level 7°F (4°C) above the temperature of the normal soil and 12°F (7°C) above that of the whitened surface. The darker soil also retained the heat gained by a few degrees at night. The emissive power of radiation of the longer wavelengths radiated by the earth is not affected by color, so the black soil did not lose heat more rapidly than the others.

Depending on the speed of growth and type of crop, the heat-absorbing effect of a darkened soil will diminish over a 4- to 10-week period as the plants grow and their leaves shade the soil. This is a harmless development, since it is only in the spring that soil temperatures need to be raised.

Plastic Mulch

The next logical step beyond charcoal dust is a simple sheet of material laid on the ground as a mulch to aid in warming the soil. Polyethylene plastic is the standard material. Plastic strips 4 to 6 feet (1.2 to 1.8 meters) wide and as long as convenient are laid on the soil with the edges buried to anchor them against the wind. Four things contribute to the popularity of plastics as a commercial mulch. Plastic mulch retains moisture, warms the soil, can prevent weed growth, and is readily adaptable to mechanization for application and removal. Both tractor-mounted and hand-pulled implements are available.

Growers have long known that clear plastic will warm the soil more effectively than black (the clear acts like a low greenhouse). However, black plastic was more commonly used since it also shades out weeds. The latest developments in infrared-transmitting (IRT) plastic mulch provide the best of both worlds. The infrared rays (which make up about 50 percent of the sun's energy) pass through to warm the soil as well as under clear plastic, but the light waves are blocked to prevent weed growth. Warm soil temperatures are vitally important for an early start with heat-loving crops. Ground-covering plastics have been available in smooth-surfaced, textured, biodegradable, and perforated styles. The perforated style allows moisture to pass through to the soil.

Low Covers

The next step beyond mulches is some sort of low covering or structure over the plants. The advantage of these simple, low structures is their flexibility.

They can be moved or erected to cover specific crops as necessary. Whereas plastic mulch used alone will justify its cost in earlier maturity of warm-season crops, in practice it is usually combined with a low plastic cover for even more improved results.

Lightweight, translucent, low plant covers have been used in horticulture for as long as the materials have been available. The problem that arises with any plant covering during the changeable weather of a typical spring is the need to ventilate the structures when the weather is too warm and close them up again as temperatures cool off. The extra labor and attention needed for such ventilation control can be a strain on the grower's resources. Recent modifications and new products have been aimed at combining cover and ventilation in one unit through slits or holes that allow the passage of air. These designs do not give as much frost protection as unperforated covers, but the difference is small, and

the trade-off benefit in the form of self-ventilation makes this idea very practical.

There are two common types of self-ventilating covering materials. The first is a clear polyethylene plastic 5 feet (1.5 meters) wide and 1½ to 2 mils thick, with slits or holes for ventilation. These are popularly known as slitted row covers. The second idea is a spun-bonded or woven translucent plastic cloth that permits the passage of air and water without the need for additional slits. These are referred to as floating covers. This is a rapidly developing field, and new products appear every year. I expect that future developments will supersede the design, but not the intent, of these covers.

The slitted row cover was traditionally laid over hoops made of No. 8 or No. 9 wire with the ends inserted in the ground on either side of the black plastic mulch. The structure stands about 16 inches (40 centimeters) high in the form of a low tunnel.

Low tunnels give protection for much less money than a greenhouse.

The floating covers are laid directly on the plants, which raise the cover as they grow. In both cases, the edges of the covering material are buried in the soil as an anchor against wind. Since these structures are low and don't easily allow for cultivation inside the cover, a weed-inhibiting plastic mulch is usually considered indispensable for weed control. For crops where a plastic mulch is inappropriate (such as early direct-seedings of carrot, radish, and spinach), the cover must be removed periodically for cultivation once the crop seeds have germinated.

Benefits. Low covers offer much more than just frost protection. In my opinion, frost protection is probably their least valuable contribution, since it amounts to only a few degrees at best. What low covers do well is create a protected microclimate beneficial to crop growth that does not otherwise exist outdoors early in the season. Low covers shelter the plant like a horizontal windbreak. They inhibit the excessive evaporation of soil moisture, allow both soil and air to warm to a more favorable temperature during the day, and maintain that improvement, albeit in a smaller way, at night. They also provide protection against insects and birds.

Spring use. Low covers can be used as early as the grower dares, depending on the crop. Since the covers provide only a few degrees of frost protection, tender crops like tomato transplants are more of a risk than hardy seed crops like carrot and radish. Many growers successfully sow extra-early carrots under floating covers, which are removed periodically for cultivation.

Removal. Covers are usually left on for four to six weeks, or until outdoor temperatures rise. Removal of the covers must be done as carefully as any hardening-off procedure. Partial removal and replacement for a few days prior to total removal is recommended so that all the advantages gained are not lost. If possible, final removal of the covers should be done on a cloudy day, ideally just before rain. Definitely avoid bright sunny periods accompanied by a drying wind, as those conditions will worsen the transition shock for the plants.

A number of years ago we combined the benefits of all of these options into a system we call Quick Hoops. We took 10-foot- (3 meter) long, ½-inch- (1.25 centimeter) diameter pieces of electrical metal tubing (EMT) and bent them around a simple form to create 6-foot- (1.8 meter) wide half circles, the ends of which could be inserted into the soil on either side of a pair of 30-inch (75 centimeter) beds with a 12-inch (30 centimeter) path between them to create a 30-inch-high low tunnel. We place these Quick Hoops 5 feet (1.5 meters) apart and cover them with 10-foot- (3 meter) wide floating row cover held in place by sandbags along the edges. Quick Hoops have been a great success for both spring and fall crops, and even in winter with a layer of clear plastic added over the row cover. We use them to winter over late-August-planted onions and late-fall-planted spinach for harvest the following spring. They have proved their worth year after year. For overwinter use we bury the edges of the plastic rather than securing them with sandbags.

In extended-season production each crop requires a decision as to whether an earlier maturity justifies the additional cost. When you are growing for a local, multiple-crop market (as opposed to a wholesale, single-crop market), it makes most sense to use covers to advance the harvest just enough for each crop to supply the demand until the earliest unprotected crops mature. Once the outdoor crops become available, the economic benefit of protected cultivation disappears. When approached within that framework, the amount of land to be covered is not excessive, even when many crops are involved.

If you want to extend the season further than is possible with Quick Hoops, you can enclose a large protected growing area effectively and economically by increasing the size of the hoops and the plastic sheet to make it a walk-in tunnel.

Caterpillar Tunnels

By "walk-in tunnels," I mean unheated structures consisting of a single layer of plastic supported by spaced arches or hoops tall enough to walk and work under. In design they vary from very lightweight units on the one hand to structures indistinguishable (except for the lack of supplementary heat) from a greenhouse. I have seen tunnels 200 feet (60 meters) long, although 100 feet (30 meters) is a more common length, and 50 feet (15 meters) may be more manageable in a climate requiring close attention to ventilation. Twelve feet (3.7 meters) is the usual minimum width and 17 feet (5.2 meters) the maximum, although the most popular models are 14 to 16 feet (4.3 to 4.9 meters) wide (enough to cover four 30-inch [75 centimeter] beds). In practice the width is a function of the materials used, the planting layout, and the range in styles, from simple to complex.

Walk-in tunnels are similar to a Quonset-shaped, bowed-pipe-frame greenhouse. The materials, however, are usually lighter and less permanent, since these tunnels are designed to be moved. Fiberglass rods, plastic pipe, metal rod, reinforcing bar, electrical conduit, and even bowed strips of wood have been used for the structure. Their ends are usually inserted into pipes driven into the ground to provide support for the arch frame. The arches are usually spaced 4 to 6 feet (1.2 to 1.8 meters) apart.

On caterpillar tunnels the plastic covering is secured in an ingenious way. The arches hold a clear plastic covering up, and tight cords are stretched over the top of the plastic from the base of the hoops on one side to the base of the hoops on the other to hold the plastic down. The plastic cover is pulled tight and attached to a T-post at either end. Ventilation is provided by raising the plastic sheet up along the bottom edge.

The friction between the arch and the cord keeps the plastic at the desired height above the ground when it is raised. Ventilation and protection, up during the day and down at night, can be accomplished quickly with a little practice. Since a larger growing area is being ventilated per unit of labor, the extra management involved is more than justifiable. Access to the tunnel is gained by raising the plastic along one edge and ducking under.

As walk-in tunnels become wider, the design changes. The ends are usually framed up separately and include a wide door. In some cases the plastic covering is secured by boards attached 3 feet (.9 meter) above the ground, along either side of the arches, and extends down to a long board lying on the ground. In this system the plastic can be rolled up for ventilation on either side by means of the long boards. At night the cover is rolled down and the board is held in place on the ground by rocks or concrete blocks.

At the upper end of walk-in tunnel design, the plastic is attached to baseboards as in a conventional pipe-frame greenhouse. The baseboards are bolted to foundation pipes spaced 4 feet (1.2 meters) apart and driven 18 inches (45 centimeters) or so into the ground.

The interior diameter of the foundation pipe is slightly larger than the outside diameter of the pipe arch. The pipe arch is erected by inserting the ends of it into the foundation pipes. The arch rests on the upper of the two bolts that secure the baseboard.

In models without roll-up sides, ventilation is provided by large doors or roof vents. The doors are framed up as large as the tunnel width permits. The roof vents can be opened and closed by hand or by automatic temperature-activated spring openers. The roof vents are an interesting construction in themselves. The desired vent sections are framed out in the roof of the tunnel, either with the same material used in the structure or with wood. The most common spacing is one vent for every four hoop sections. The plastic covering is secured around the edges of the roof opening, most easily with a two-part channel and wire fastener attached to the edges of the vent opening to hold the plastic. The two parts snap together and provide a strong, reliable grip on the edge of the plastic. A lightweight frame for the roof vent hatch is covered with plastic, hinged at one end, and attached to the temperature-activated opening arm.

With any plastic structure, even when conventionally covered, wind whipping and abrasion can be a serious problem. No matter how carefully the cover is tightened when it is first put on, it always seems to loosen. There are two ways to deal with this. The simplest is to run stretch cord over the top of the plastic from one side to the other, as with the caterpillar tunnels. One cord between every fourth rib is usually sufficient. The tension of the stretch cord will compensate for the expansion and contraction of the plastic due to temperature changes and will keep the cover taut at all times. The second solution is to cover the tunnel on a very warm day when the plastic is expanded. Subsequent cooler temperatures will keep it tight. Or for greatest security cover the

house with two layers of plastic and inflate the space between them with a small squirrel cage fan. This creates a taut outer surface that resists wind and helps shed snow.

The final step beyond these advanced walk-in tunnels is to add a source of heat other than the sun. At this point you have a greenhouse, so it might be better to deliberately build one from the start.

Greenhouses

The simplest greenhouse is only a stronger and slightly more complicated walk-in tunnel. The major differences are that the greenhouse usually has greater structural stability and provides supplementary heat. A greenhouse can be built of 2×4s and will look very similar to a house frame. Or it can be constructed in the same bowed-pipe Quonset design as the walk-in tunnels using a heavier-gauge tubing. I

My first homemade greenhouse in 1972.

prefer the latter style. Bowed-pipe arch greenhouses can be home-built or purchased in modular units that are simple to erect, lengthen, or move as desired. Pipe, being less massive and less numerous than wooden structural members, casts almost no shadows and allows a maximum amount of light to enter.

The two layers of air-inflated plastic mentioned above are commonly used for greenhouses. This air-inflation idea adds strength and stability and makes the greenhouse more rigid. The outer plastic skin remains tight in all temperatures and is not subject to the flapping, abrasion, and tearing that single layers experience during windy weather. Furthermore, the stressed curve of the surface sheds snow better and, except in the heaviest storms, it will not need to be swept or shoveled off. When inflated, the two layers are separated by about 4 inches (10 centimeters) of air, providing an insulation layer that results in a 25 to 35 percent savings in heating costs

compared with an identical greenhouse covered with a single layer.

Supplementary heat for the greenhouse can be provided by burning any number of fuels, but the most popular small greenhouse heaters, and the ones I recommend, burn natural gas or propane. They come equipped with a fan blower and are vented to the outside. Installation is relatively simple, and the units can be moved along with the greenhouse.

Energy is an important issue. Some growers may want to explore the greenhouse heating potential of wood, decomposing organic matter, passive solar storage, or some other fuel. I heartily concur with this concern. All I suggest is that you postpone solving the energy problem for a few years. My goal is to present a farming system that works, and you must decide at the outset where your priorities lie. Is this a horticultural unit or a forum for experimenting with new energy technologies? For beginners it is always

Bending pipe for today's homemade greenhouses.

a wise practice to use a dependable conventional technology, because most of the problems have already been worked out by others.

Along with the provision of heat comes the need to mechanize the ventilation system. Thermostatically controlled fans and automatic shutters do the best job and are conventionally used in greenhouse temperature-management systems. Two-stage thermostats allow for different quantities of air to be moved, depending on temperature demands, and will prove most effective in precisely regulating the ventilation system.

Another piece of advice is appropriate here. Although the idea of two-stage thermostats, automatic shutters, and ventilation fans may seem overly technological and complicated, they emphatically are not. A greenhouse structure can quickly overheat to temperatures that are deadly to plants. A depend-

able ventilation system is crucial. Occasionally, such mechanisms may malfunction, and yes, they are more expensive than hand-operated vents or roll-up sides. But they are parts of a successful system, and that system works reasonably well. Any questions should wait until you have gained some experience with greenhouses. Once you are involved with greenhouse growing, you'll see the system makes sense. The rest of the vegetable farm will be demanding enough that you will appreciate having some automatic assistants to look after a few details.

COMMERCIAL GREENHOUSES

For commercial vegetable production, a 17- to 30-foot- (5.2 to 9 meter) wide, 50- to 100-foot- (15 to 30 meter) long bowed-pipe arch model covered with either a single layer or double-air-inflated polyethylene plastic best meets the needs of the small-scale producer.

End-wall ventilation.

The smaller houses are an efficient size to grow in and are manageable enough to disassemble and move if desired. Wider houses are even more convenient, as long as they fit a grower's production system. Two points will need to be considered. First, in a multicrop-production system there is greater flexibility with two or more smaller houses than with one large one. Since there will be different crops grown, each requiring a different temperature regime, a whole house can be given over to one crop or a group of crops with similar cultural needs. Second, if something does go wrong, it will threaten only a part of the production rather than all of it.[2]

GROWING IN A GREENHOUSE

The aim of a greenhouse is to create climatic conditions that are optimal for plant growth. The same should apply to soil conditions. Plant growth is intensified in the greenhouse, so extra care must be taken with soil preparation. All the factors that were stressed in the chapter on soil fertility are triply stressed here. Applications of well-finished compost are required for successful greenhouse culture. Greenhouse tomatoes and cucumbers are two of the most demanding crops; they need exceptional fertility, not only for soil preparation but also as a compost top-dressing applied once every six weeks. Lettuce, the other major greenhouse crop, will do well in a rotation with either tomatoes or cucumbers without further soil amendment.

Heated greenhouse production of high-demand crops is a valuable magnet to attract customers to the rest of the farm's produce. It can also be a highly lucrative area of specialization in its own right. But don't jump into it without first giving the matter considerable thought. Growing high-quality greenhouse vegetables demands a real commitment to management and an attention to detail. Serious commercial-greenhouse production should be postponed until the farm is well established and more capital can be acquired. Then I would recommend it. Books with detailed information about greenhouse vegetable production are listed in the bibliography.

Tomorrow's Answers

The progressive grower always keeps one eye on doing a job and the other on how it could be done better. You should always be alert to improve both the practical and economical aspects of your present production technologies, not in order to follow the future, but rather to lead it. The future can often be predicted based on today's patterns of development. The trend in season extension is quite clear. The pattern runs in the direction of making the structures lighter, simpler to manage, and more mobile.

Today's ideas were yesterday's suggestions. Tomorrow's ideas will rise from today's problems. The best new concepts come from experienced practitioners and were devised to answer their needs. Once you gain experience, the next improvement for you as a small grower, and for countless other food producers, is just around the corner.

CHAPTER TWENTY-FIVE

The Movable Feast

The ideal greenhouse system for organic growing would be as forward thinking in terms of "natural" protected cultivation of vegetables as the leading-edge developments of organic growers over the years have been in terms of "natural" soil fertility. The temporary plastic-covered caterpillar tunnels come close. After covering the ground when needed, they can be disassembled and moved to another location. They can be preceded or followed by a green manure. They avoid difficulties like pest and disease buildup and the soil nutrient excesses often encountered with permanent greenhouses. It would be nice to have that periodic outdoor exposure of the soil allowed by temporary structures, in addition to the benefits of a

Close planting and successional sowing make full use of greenhouse space.

full-function, permanent greenhouse. Fortunately, that combination is possible. The answer is a greenhouse that can be moved without disassembling.

The Mobile Greenhouse

A mobile greenhouse, by which I mean a fully equipped greenhouse that can slide or roll to one or more alternative sites, combines the climatic benefits of a permanent greenhouse with the soil-cleansing benefits of a temporary tunnel. The idea of a greenhouse that could be moved without being dismantled became a reality at the end of the 19th century. Its development was a response to the search for a better solution to greenhouse soil sickness. The options at that time—removing and replacing the soil to a depth of 12 to 16 inches (30 to 40 centimeters) or sterilizing the soil with steam—were considered less than satisfactory because of high labor costs or the soil fertility problems following sterilization.

The early mobile greenhouses of 100 years ago were quite a piece of work. I'm sure the idea of building something as fragile as a glass greenhouse to be strong enough to move from one site to another without self-destructing posed a challenge to the original innovators. Nevertheless they persevered. The houses they built looked like conventional greenhouses on the surface—a wood or steel frame that was covered with glass. But instead of being attached directly to a foundation, the frame of the mobile house was supported by metal wheels that rolled on metal rails. The rails were solidly attached to a conventional foundation. The end walls were planned so the bottom part could be raised to pass above crops when the house was rolled to one or more alternative sites.

Some models took a different approach. They were designed so only the roof moved. In those cases the roof trusses made a stable triangular structure that rolled on rails set on top of the side walls. That style offered extra value in a windy location. Even when the roof was uncovered, the walls alone served

as a translucent windbreak and provided considerable climate enhancement. Despite the initially higher costs of mobile greenhouse designs, many innovative growers thought this a worthwhile experiment and quite a few were built.[1]

I saw my first mobile greenhouse while visiting an organic market garden in Holland during the early spring of 1976. Although newly constructed, it was just like the classic models I had read about. The metal frame covered with glass was 40 feet (12 meters) wide by 120 feet (36 meters) long. It sat on metal wheels running on railroad rails long enough so three different sites could be served. The morning I arrived the house was being moved. Even though I was familiar with the idea, I will admit my surprise at seeing such a huge structure trundling slowly toward me, pushed by workers on each rear corner. The irrigation, heat, and electricity inputs had been disconnected prior to moving and then were reconnected to new outlets at the second site. The early crop of lettuce that the house had been covering was now advanced enough to finish its growth outdoors. Tomato transplants, ready to be set out in the newly protected site that afternoon, were waiting alongside.

I was told that, except for certain special situations, such as out-of-season strawberry and flower crops, very few mobile houses were still being built. Most Dutch growers had shifted to solving their greenhouse soil problems with chemicals or by growing in artificial media. However, my hosts believed the mobile technology made great biological sense for their organic operation. The same reasons that inspired its initial development—preventing pest buildup and soil fertility problems—still applied to those growers using natural methods. And although in this case the cost averaged about 20 percent higher than a static glass house, the returns were also 20 percent higher. My hosts also stressed another virtue of mobility. The capital cost of the greenhouse could be recovered more quickly because it could effectively cover more crops in a year than could be protected by a stationary greenhouse.

Four moveable frame units covered as one.

A sliding-design movable greenhouse

The Advantages of a Mobile Greenhouse

First, I want the same thing that the early mobile greenhouse pioneers wanted—more natural soil conditions. By exposing the soil to the outdoors—rain, wind, snow, direct sunlight, and serious subfreezing temperatures—I can bring into play many natural cleansing and balancing processes that are denied to a permanently covered, full-function greenhouse. Second, and this point expands on the first, mobile houses permit me to use a wider crop rotation. I can now include long-term green manures without sacrificing any greenhouse cropping potential, because I grow them on an uncovered site. Soil-improving crops are a definite plus, not only for their physical and biological effects on the soil, but also to reduce the amount of compost that needs to be brought into the greenhouse. The greenhouse soil can now become as self-supporting in fertility as

the rest of the farm (see chapter 13). Third, I find that mobility allows me to avoid the more artificial and palliative-centered approaches (for example, steam sterilization, grafting of tomatoes to resistant rootstocks, constant pest monitoring, trapping, and supplementation of beneficial insects) to which other organic greenhouse growers are turning as the problems of continuous greenhouse cultivation intensify year after year. I readily admit to a lack of enthusiasm for palliatives. I prefer to look for ways to encourage natural processes to do what they do so well if I just let them do it.

And fourth, summer greenhouse crops like tomatoes can be left growing as late into the fall as temperatures allow since the winter crops have already been sown on an adjacent site at the appropriate date. The greenhouse is not moved to cover the winter crops until they need protection.

A fifth advantage is the saving of money and energy for both heating and cooling. For example, I

Winter greenhouses need good perimeter ditches for drainage.

202

start planting many winter harvest crops in midsummer and continue on through early fall. These winter crops are planted on the uncovered site, where they grow vigorously until the shortening days and cooler temperatures of late fall slow their growth. I then move the greenhouse over them to provide protection for winter harvesting. If I had started these crops in a stationary greenhouse, I would have needed an expensive cooling system. Even with the vents open and doors removed to provide as much passive cooling as possible, the summer greenhouse is too warm for starting winter crops. But the summer greenhouse is an ideal place to grow heat-loving crops. And so I do.

The only cost for all these biological advantages is the additional expense of making a static house mobile. If that additional expense is reasonable, the idea becomes very attractive. So when I became serious about building a real mobile greenhouse, I began by checking the prices of the mobile plastic greenhouses offered by European manufacturers. Unlike the economic figures given me by my Dutch friends years ago, these new houses are no longer 20 percent more expensive. I received a quote from one manufacturer that was triple the initial price of the house. Of course a lot of that price differential relates to the difference in initial cost between glass and plastic houses. But it still seemed way out of line. My goal was to find a mobile greenhouse technology that would add no more than 10 percent to the cost of a plastic greenhouse, and yet would be strong enough to be used on a 30-by-96-foot (9 by 29 meter) house. Further, since I already owned stationary plastic houses, I wanted a system that could make them mobile without too many added parts. During the occasional frustrations that inevitably occur while trying to develop any new idea, I asked myself more than once, "Do I really need mobile houses? Can't I do just as well with ones that don't move?" Every time, one after another of the reasons stated above would come up as a justification to continue my efforts. I'm glad I did.

Overhead irrigation in a movable greenhouse.

However, there is a lot to be learned from other people's experience. A large farm I am acquainted with, which enthusiastically built a couple of in-line 30-by-100-foot (9 by 30 meter) movable greenhouses, changed them to a non-movable by adding extra ribs to end up with a 30-by-400-foot (9 by 122 meter) house. They made that decision for the greater efficiency of using larger-scale mechanical equipment for tillage, planting, and harvesting the greenhouse crops. However, since they felt that a soil-improving cover crop was truly needed for soil health, they now take the plastic off every three years to overwinter the house in its natural environment with a cover crop. There is more than one way to reap the movable greenhouse benefit of a temporarily uncovered soil.

The Designs

My first mobile greenhouse design years ago had a wooden rail attached to pipes driven into the ground. The wooden understructure of the greenhouse could slide along that wooden rail with the help of some liquid soap. That prototype was our proof of concept and convinced us of the value of movable houses. We continued experimenting with the goal of making the mobility process simpler and less expensive. In the next design the pipes, which were driven into the ground along each edge of the greenhouse site, had a large ball-bearing caster set on top over which the angle-iron base of the greenhouse could roll. Two people could move our 20-foot- (6 meter) wide by 36-foot- (11 meter) long prototype. We employed that design to build our first commercial-scale movable greenhouses, 100 feet (30 meters) long by 30 feet (9 meters) wide. These early houses had only two sites. Water and power were supplied underground to a 2-foot- (60 centimeter) wide strip that was common to both sites. Our 26-horsepower tractor could easily move them. As with all of our movable greenhouses, they were chained to 4,000-pound- (1,815 kilogram) test ground anchors when they were in place.

There were a few technical problems with the ball-caster system, and it was a lot of work to drive the support pipes into our rocky soil, so we changed to a sliding sled concept, using metal runners as the base of the greenhouse for our next design. That solved some problems, but the friction of the runner on the soil required more power to pull those models, and we had to hire heavy equipment. At this point we also decided that the 100-foot-long (30 meter) houses created difficulties without any commensurate benefits. All of our mobiles since then have been 50 feet (15 meters) long. To make the houses easier to move, the next design used metal wheels (one on the bottom of each hoop leg) rolling on pipe rails lying on the ground. This was a nice design but had two problems. First, we had a lot of money tied up in all those wheels sitting under each greenhouse. In addition, as with any movable system whether using ball bearings on vertical pipes or wheels on horizontal pipes, the greenhouse always sits slightly above the ground. That gap must be sealed to keep cold air out and then it must be unsealed before every move. Another version of the wheel-on-rail model, which we designed for an American greenhouse company that presently sells them, uses an 8-inch- (20 centimeter) wide V-track instead of the pipe rail and rolls on heavy-duty equipment wheels with excellent ball bearings. It moves more easily than any of our other designs, but given the cost of the high-quality track and wheels, it is much more expensive.

While designing greenhouse structures and devising better ways to move them, we were also redesigning the end walls. Our original polycarbonate end walls with hinged horizontal panels along the bottom and a removable door were expensive and awkward. So we began covering the end walls with plastic. Instead of a removable door, we built the end wall with a crossbar all the way across at a height that would clear the top of the crops. We had to step over that crossbar to enter the greenhouse, but that was only a minor inconvenience. For more detailed information, please see chapter 10 in *The Winter Harvest Handbook*.

Our very latest movable greenhouse design includes all the desirable features we have arrived at over the years. The greenhouse is 50 feet (15 meters) long and 30 feet (9 meters) wide. The end wall can be opened at the bottom to pass over our low tunnels, the entrance door can be put in the side wall, and the bottoms of the hoops on each side are attached to a square metal base pipe that sits flat on the ground. For moving we attach five wheelbarrow wheels with pneumatic tires to the metal base pipe on each side of the greenhouse. An axle on the "lift wheel" is placed into a hole in the metal base pipe while the greenhouse sits on the ground. The chains and chain binders are removed. Then a vertical lever that is part of the lift wheel is pushed down and pinned in place, effectively lifting the greenhouse a few inches off the ground. After moving, the wheel units are unpinned, removed, and stored under cover. In this way the same set of wheels can now be used for every greenhouse on the farm. The house rolls easily enough to be pulled with a hand winch, thus no longer requiring tractor power. We are presently changing over all our mobile greenhouses to this new system. It is as close to a perfect design as we have come up with yet, but I am sure we will be making more changes in the future. Even when I finally decide on a design, I always know it can be improved. I believe in a simple law and corollary. The law states, "There is always a way." The corollary states, "Once you find a way, there is always a better way."

CHAPTER TWENTY-SIX

The Winter Harvest Project

Part of my fascination with greenhouses and greenhouse systems arises from a desire to supply food to my customers for as much of the year as possible. I think I benefit their health by growing vegetables of an exceptional quality. I know I help support a vibrant local economy by keeping the money circulating within my community. And as I mentioned in the marketing chapter, as a local grower I have an important advantage over the large wholesale shipper: I provide fresh vegetables. So I don't care if the trucked-in crops can claim they were grown with pure stardust by elves and fairies. Any vegetables picked in bulk and shipped through the wholesale system are not fresh. Mine were picked today and are on the customer's table tonight or tomorrow. I can compete on that fact alone.

If I'm going to stress that freshness angle as part and parcel of my farm, I want to make it happen for as much of the year as possible so as not to disappoint my customers—whether stores, restaurants, co-ops, or neighbors—or lose them to another supplier. But I want to do this without excessive expenditure of energy. The winter harvest succeeds by combining the technology of climate modification with the biology of the vegetable world. The technology consists of two layers of protection to temper the harsh blasts of winter; the plants grow under a layer of spun bonded fabric inside an unheated greenhouse. The biology involves the selection of cold-hardy vegetable varieties. These traditional winter crops don't mind the short days,

nor are they harmed by freezing or by remaining frozen for periods of time.

Our goal has always been to extend the fresh vegetable harvest season throughout the winter months by using low-input solutions powered by on-farm resources. The criteria for achieving that goal focus on environmentally sound practices that minimize use of energy and resources. The list of criteria is

An inner layer of floating row cover.

205

posted on the wall of the greenhouse to keep us paying attention to how we might do it even better:

Climate-adapted. Exploiting the potential of simple, protected microclimates for the production of hardy crops during the cold months.

Sunlight-driven. Minimal technology pipe-frame greenhouses covered with a single layer of exterior plastic and adding single-layer interior covers. No supplemental heat, no perimeter insulation.

Nature-directed. Working with rather than against the realities of cold conditions. Successfully growing cold-hardy crops in the cold months through understanding the influence of day length and soil temperature on time from planting to harvest. Benefiting from the adaptation of numerous cultivars to winter light and winter temperatures.

Farm-generated. Using homemade compost, cover crops, and sod-based crop rotations as the principal inputs to nurture a fertile soil.

Claytonia.

Arugula.

Hakurei turnips.

Komatsuna.

Pak choi.

Harvesting mâche by cutting at soil level.

Tokyo Bekana.

Baby beet leaves.

Western Front kale.

Pest prevention. Emphasizing balanced soils as part of a plant-positive management system. Including mobile greenhouses to avoid pest buildup in covered environments.

Weed prevention. Focusing on shallow tillage, timely cultivation, solarization, and flame weeding to prevent weeds from growing or going to seed.

Practical tools. Searching out the simplest, most effective tools (and often making or modifying them ourselves) for each task from soil preparation through harvest.

Nutritionally sound. Providing all the nutritional benefits of truly fresh food by focusing on local production for local markets.

Resource efficiency. Saving energy since winter crops locally grown in unheated structures require only 5 percent of the energy used for long-distance transport.

Eco-rationality. Stressing both *economic* and *ecological* principles. Earning a good living for farmers and having a positive influence on the health of our customers and the local environment.

In order to achieve a decent income during the coldest months of winter we expend some energy by adding minimal heat to one greenhouse. We keep it just above freezing by burning wood and propane. That gives us a greater variety of crops, beyond those that are successful in the unheated houses, to attract customers to our winter farmers markets. The increased business offsets the cost of heating by about four to one.

Weeds

My weed solution in the winter greenhouse is to dispatch them young and never let any go to seed. I can't emphasize this point enough: Get on top of weeds at the start, and stick to it. The way to achieve that goal is very straightforward. Most weed seeds germinate in the top 2 inches (5 centimeters) of the soil. Get rid of those seeds, and the problem is solved. As long as

I till shallowly and don't continually bring up new seeds from lower layers, the weed seed reservoir in the top 2 inches of soil is exhausted after three to five years, and the foundation for clean cultivation has been laid.[1] So for weed-control purposes I think of the winter greenhouse as the "Five Centimeter Farm" and use two-weed prevention / weed-suppression techniques. First, I do all the surface soil preparation with the electrically powered Tilther, which is designed to only till shallowly and let deep weed seeds remain buried. Second, we solarize the greenhouse soil for a month in summer every three years, as necessary, to kill any weed seeds in the top 5 centimeters by reaching a soil temperature of 145°F (63°C) at the 2-inch depth.

Pests

I have experienced very few insect and disease problems in the winter garden. Initially I felt the greenhouse might be introducing an artificial factor that could make plants more susceptible. I am convinced now that any effect is minimal as long as well-finished compost and crop rotations are used. The only disease I have seen was a little gray mold fungus (*Botrytis* spp.), which didn't surprise me, since it thrives in cool, damp conditions. I now prevent it by venting off moist air in the mornings. Prevention is a time-honored greenhouse practice. In Dutch organic greenhouses I have seen specially designed vacuum cleaners used before planting a new crop to remove any leaves or other partially decayed organic residues from the soil surface so they cannot serve as substrates for disease organisms. I don't use a vacuum, but I do spend time cleaning up crop residues and keeping stems trimmed close. I try to manage the greenhouse environment so as to encourage only the plants.

European growers have a nice trick for improving airflow to prevent problems in midwinter lettuce crops. Instead of setting the soil-block-grown transplant seedlings into the soil at the usual depth, they set them on top of the soil. With a little irrigation the plants will root vigorously from the bottom of the block. Depending on block size, this keeps the lower leaves an inch or more above the soil surface, providing better air movement and less chance of bottom rot.

Fall and winter freezing may have a lot to do with the absence of pests. I have asked entomologists for information on that subject. I'm curious to know whether anyone has researched the use of selective low temperatures as a pest-control measure. I am aware that many bedding-plant growers turn off their houses in the winter to freeze out pests, but I have never heard of anyone doing it while plants were growing in the house. The idea would be to take advantage of any differential hardiness between plant and pest by allowing the temperature on a freezing night to drop to a point where the pests were killed but the plants survived.

A few entomologists, after I explained the parameters of our system—mobile greenhouses, compost fertilization, crop rotation, green manures, winter crops—commented that, given what they knew about insect life stages and reaction to low temperatures, they wouldn't be surprised not to find any pest problems at all. I found their comments reassuring, and definitely in line with my practical experience. For whatever reason, pest control is a very minor

Harvesting winter spinach.

concern for the winter harvest crops. Which is nice, because it allows me to concentrate on the most important consideration—crop quality.

Crop Quality

The quality of winter garden crops is continually high, both visually and in terms of flavor, tenderness, and just plain eating pleasure. We do everything we can to ensure optimal soil conditions, and it obviously pays off.

The major quality problem sometimes seen in winter hothouse crops—high nitrate content—is not a problem in this system. That's because none of the causative factors are present: use of soluble nitrogen fertilizers, forcing crops in low-light conditions at high temperatures, lack of trace elements, and the use of susceptible varieties. We fertilize only with well-humified, one-and-a-half- to two-year-old compost; we don't force the crops; we have plenty of trace elements in the soil, and we are not growing special winter-forcing varieties.

Plus there are positive factors operating, thanks to the mobile greenhouse. There is no continuous buildup of nitrates or other salts in the soil as there would be in a permanent greenhouse, because this soil is uncovered for one year out of every two. In addition, the crops that we are harvesting during the lowest winter light (mid-November to mid-February) were all sown outdoors in the fall. They stopped their growth naturally as a result of declining temperatures and day length. The greenhouse covering then arrived to protect them (the same as if we had moved them to a milder climate). We have extended the harvest season for crops sown during the growing season.

The high quality that can be achieved with winter garden crops is not solely attributable to organic methods. I have seen problems on European farms that used dried blood and feather meal fertilizers to force winter crops. Blood and feather meal are just as likely as soluble chemical nitrogen to result in high nitrate levels in winter produce. High-quality winter crops are a result of understanding the constraints of winter production and taking care to create optimal, balanced conditions for plant development.

Low Winter Sun and Greenhouse Layout

The other reality of short winter days is low sun angle. Be sure your greenhouse site won't be shaded by trees, hedges, houses, a mountain, or other greenhouses. The amount of change in angle from summer sun to winter sun always surprises me. At our latitude (44 degrees north) the noon sun has an altitude angle of 69 degrees on June 21, but only 22 degrees on December 21. I have to constantly remind myself to pay attention to where the winter shadows will fall when I am laying out greenhouses.

I plan my greenhouses with the long axis running east-west to maximize winter sun. To the south of 40 degrees north latitude, a north-south orientation is recommended. You need to be aware that true south for greenhouse layout is not the same as magnetic south at most locations. You will need to determine your magnetic declination from a topographic map and adjust accordingly.

When the sun is low in the winter sky, a 10-foot- (3 meter) tall greenhouse running east-west will cast a shadow 25 feet (7.6 meters) long; a 12-foot- (3.7 meter) tall greenhouse will cast a 30-foot (9 meter) shadow. Thus, a second greenhouse behind the first should be sited at least that far away so as not to be shaded. In the past I have often used that winter-shade area to grow outdoor summer crops (they benefit from the windbreak created by the greenhouses), or I have covered that ground with temporary greenhouses in spring and summer when the sun is higher in the sky.

Another option is to take advantage of movable greenhouses and plan the rotation so the greenhouses are offset. That way you can place them as close together as is convenient, yet still avoid winter shadows. The sample rotations shown in figure 26.1

are planned so the winter positions will alternate between sites A and B.

There is one final reality of winter in addition to cold temperatures and low sun: snow. With heated greenhouses the standard practice is to turn up the heat during a snowstorm so as to melt off the snow. For unheated and low-temperature production, I take a different approach. My greenhouses have a Gothic-arch roofline and heavier pipe construction; both features make for a stronger structure with a shape that sheds snow better. Although the snow slides off the roof, it doesn't go away. A heavy snowfall remains piled up against the sides and can block sun. If you shovel by hand, pay attention. You can easily jab the snow shovel through the greenhouse plastic (I have done it more than once). If you have a lot of greenhouses, a snow blower is the tool of choice. The snow blower can remove snow from

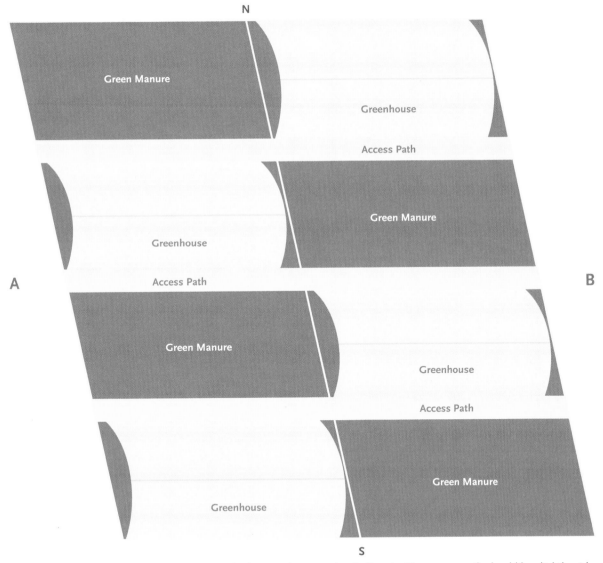

Figure 26.1. Siting greenhouses to avoid cast shadows and to provide windbreaks. The access path should be slightly wider than the height of the greenhouse.

right alongside the wall of the greenhouse without creating back pressure as a snowplow would.

Note: For more detailed winter growing information, I encourage market growers to consult my book *The Winter Harvest Handbook: Year-Round Vegetable Production Using Deep-Organic Techniques* *and Unheated Greenhouses* (Chelsea Green Publishing, 2009). It includes extensive and up-to-date information regarding use of row covers in unheated greenhouses, the rationale for minimally heated greenhouses, tables of temperature survivability of various greens, and a crop list with a seeding schedule.

CHAPTER TWENTY-SEVEN

Livestock

One way to diversify your farm's potential marketing options and simultaneously improve soil fertility for vegetable crops is to include livestock in your system. Because adding animals to the farm can increase the management load considerably, however, you should not do so until your basic vegetable-growing operation is firmly established. Even then, a lot of thought must be given to the added hours of work and the possible need for extra workers.

If you feel comfortable taking on these new responsibilities, the addition of livestock is the next logical step in enhancing your farm's stability and economic independence. Livestock can be considered as a part of the crop-rotation plan or as a separate operation altogether.

Free Manure

Many vegetable growers have raised livestock separately from their crop rotation, using mainly purchased feed for their farm animals. The stock were not usually kept for profit, but rather as a source of free manure. A delightful old book published in 1864, *Ten Acres Enough* by Edmund Morris, details the author's positive experience in doing just that. For example, if the farm has barn space, you might decide to raise dairy bull calves to beef market weight on purchased hay and grain. As a supplementary farm activity, this is not likely to yield a profit per se. The costs will likely equal the income from selling the animals. But the profitability must be calculated from the savings gained by not having to purchase manure, and the further benefit of having the quantity and quality of manure you need produced right on the farm.

Horses

From my experience with different manures as fertilizers for vegetable growing, I would recommend horses over cattle. If the facilities were available, I would choose to board horses for the winter. In many ways this may be a simpler option. First of all, winter is a slack time in the vegetable grower's year, and the livestock responsibility would not be continued through the busier half of the year. Second, I could charge enough to feed the horses well and bed them on straw to produce a high-quality manure-straw mixture. Lastly, even if my return from the operation were only enough to cover expenses (I could thereby underbid other horse boarders to get the number of animals I desired), I would have produced, at no cost, a year's supply of what was long considered to be the ideal soil amendment for general vegetable crops: horse manure and straw bedding. This fibrous horse manure–straw combination has been a reliable fertilizer throughout the history of market gardening.

The size of the livestock operation can be calculated according to the farm's manure requirements. In order to manure half the acreage every year at the rate of 20 tons per acre (18,145 kilograms/4,000

213

square meters), a 2-acre operation would need 20 tons of manure. Since a horse will produce 15 tons (13,610 kilograms) of manure (with bedding) per year, that would equal 0.066 horse per ton. For a six-month boarding operation, that factor must be doubled to 0.133; 0.133 horse per ton of manure × 20 tons = 2.65 (call it 3) horses. Boarding three horses bedded on straw for six months would give you 20 tons of first-class vegetable fertilizer. Mix that half and half with plant waste when composting, and you would have a truly superior product.

Managing the Manure

There would be one problem to contend with, however. On the scale of production we are considering, the grower would be faced with a bit of work managing that manure properly (by composting it) and spreading it on the field. However, if the stalls were cleaned every day and the manure added to a steadily growing compost windrow, the composting part would be under control. Spreading that quantity of manure is a more formidable task. I have spread 20 tons (18,145 kg) of manure by hand in a year, and I've done it for many years of vegetable growing. Yes, it is hard work, but certainly not beyond the ability of most people. It is usually accomplished over a period of time, and in retrospect it is not all that difficult. I will agree that spreading 20 tons of manure would be a lot easier with some machinery. So once we had the farm on a firm financial basis (and were also a number of years older), we purchased a small tractor with a front loader and that has made our work a lot easier.

Rotating Livestock and Crops

As I said, were I to operate such an animal manure program, I would choose horses. But other considerations convince me that the first option—including livestock in the rotation—may be a better solution for the following reasons:

- Small livestock products such as fresh eggs can be valuable as a means of attracting and keeping customers for the vegetable operation. The livestock / soil-fertility combination will then contribute directly to farm income.
- The ideal soil structure and organic-matter benefits conferred by adding manure to the soil can also be achieved by growing a mixed legume-grass sod and grazing it with livestock.

Fortunately, there is a livestock choice that will make optimal use of a legume-grass sod, provide a readily salable item, not require much extra care, and effectively produce manure and spread it for you in the process.

Poultry

The best livestock to complement vegetable production are poultry ranged on sod in the rotation. Chickens thrive when run on shortgrass pasture, known as range. According to varying experience, pasture can provide up to 40 percent of their food needs. In this option poultry are grazed on a grass/legume pasture that is included in the crop rotation. In that way the grass/legume crop grown for soil improvement also feeds the livestock, and they, in turn, manure the field.

Studies show that a grazed pasture gives higher crop yields (when it is plowed up for arable crops) than an identical pasture that was mowed with the clippings left in place. There is a significant soil-fertility benefit from the biological activity of animal manure, even though some of its ingredients came from the field itself and some nutrients were actually removed by the livestock. Since laying hens are supplemented by feeds purchased off the farm, the fertility gain will be significant. On average, 75 percent of the fertilizer value of the feed consumed is returned to the soil in manure.

The first requirement in this range system is for some sort of movable poultry house.

Close-up of the roll-out-style community laying box.

The Chickshaw

There are many ways to house poultry on range. Range-rearing systems were used extensively prior to the 1950s, and many styles of shelter were devised.[1] When we began with range poultry back in 1965, we modified those early designs to make the houses smaller and lighter so they could be moved without a tractor. We also built them with 26-inch (65 centimeter) cart wheels for easier rolling. This design was instantly christened "the Chickshaw."

Although initially made as a small wooden house, we now make the superstructure of the Chickshaw from curved lengths of 1-inch (2.5 centimeter) EMT bent around the same form we use for Quick Hoops (see chapter 24). That creates a round roof that we cover with a layer of thin fiberglass or other opaque

bendable material. The base frame is made from lengths of 1⅜-inch- (3.5 centimeter) diameter chain-link-fence top-rail connected by T-braces. The floor is made from 1-inch (2.5 centimeter) wire mesh. Half-inch (1.25 centimeter) hardware cloth covers the end walls. A wooden door hinged at the bottom serves as an entry ramp for the chickens and also gives us access to the interior.

We start young poultry in the Chickshaw from the day they arrive. The 5-foot-by-6-foot (1.5 meter by 1.8 meter) dimensions are large enough to house 100 baby chicks in each Chickshaw at the start. When the Chickshaw is used for brooding chicks, the wire mesh floor is lined with newspaper beneath a layer of wood shavings. We hang two electric heat lamps from the ceiling inside and adjust their height above the floor to achieve the ideal temperature.

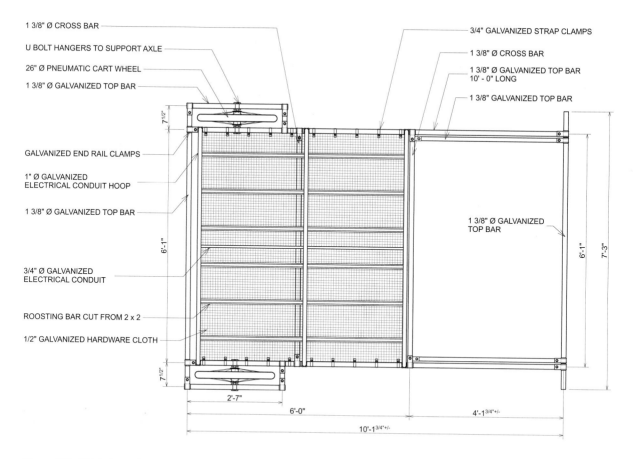

1 3/8" Ø CROSS BAR

U BOLT HANGERS TO SUPPORT AXLE

26" Ø PNEUMATIC CART WHEEL

1 3/8" Ø GALVANIZED TOP BAR

GALVANIZED END RAIL CLAMPS

1" Ø GALVANIZED
ELECTRICAL CONDUIT HOOP

1 3/8" Ø GALVANIZED TOP BAR

3/4" Ø GALVANIZED
ELECTRICAL CONDUIT

ROOSTING BAR CUT FROM 2 x 2

1/2" GALVANIZED HARDWARE CLOTH

3/4" GALVANIZED STRAP CLAMPS

1 3/8" Ø CROSS BAR

1 3/8" Ø GALVANIZED TOP BAR
10' - 0" LONG

1 3/8" GALVANIZED TOP BAR

1 3/8" Ø GALVANIZED
TOP BAR

7 1/2"

6'-1"

7 1/2"

6'-1"

7'-3"

2'-7"

6'-0"

4'-1 3/4"+/-

10'-1 3/4"+/-

Figure 27.1. Chickshaw plan.

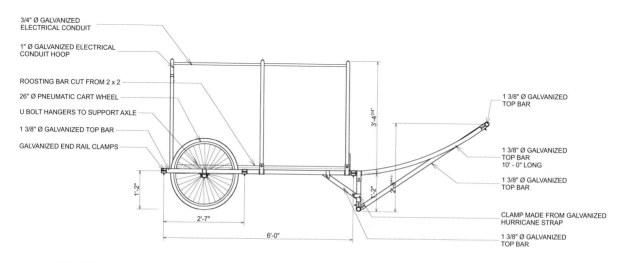

3/4" Ø GALVANIZED
ELECTRICAL CONDUIT

1" Ø GALVANIZED ELECTRICAL
CONDUIT HOOP

ROOSTING BAR CUT FROM 2 x 2

26" Ø PNEUMATIC CART WHEEL

U BOLT HANGERS TO SUPPORT AXLE

1 3/8" Ø GALVANIZED TOP BAR

GALVANIZED END RAIL CLAMPS

1 3/8" Ø GALVANIZED
TOP BAR

1 3/8" Ø GALVANIZED
TOP BAR
10' - 0" LONG

1 3/8" Ø GALVANIZED
TOP BAR

CLAMP MADE FROM GALVANIZED
HURRICANE STRAP

1 3/8" Ø GALVANIZED
TOP BAR

3'-4 3/4"

1'-2"

2'-7"

6'-0"

Figure 27.2. Chickshaw side elevation

Each Chickshaw has a roll-out-style community laying box.

CONNECTION DETAIL

3" = 1' - 0" (QUARTER SIZE)

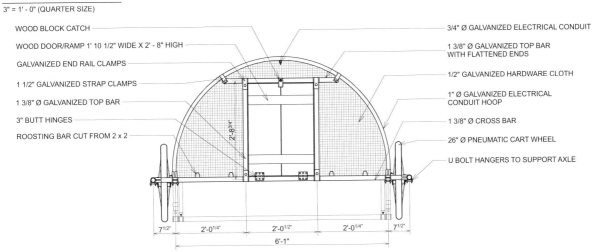

WOOD BLOCK CATCH

WOOD DOOR/RAMP 1' 10 1/2" WIDE X 2' - 8" HIGH

GALVANIZED END RAIL CLAMPS

1 1/2" GALVANIZED STRAP CLAMPS

1 3/8" Ø GALVANIZED TOP BAR

3" BUTT HINGES

ROOSTING BAR CUT FROM 2 x 2

3/4" Ø GALVANIZED ELECTRICAL CONDUIT

1 3/8" Ø GALVANIZED TOP BAR WITH FLATTENED ENDS

1/2" GALVANIZED HARDWARE CLOTH

1" Ø GALVANIZED ELECTRICAL CONDUIT HOOP

1 3/8" Ø CROSS BAR

26" Ø PNEUMATIC CART WHEEL

U BOLT HANGERS TO SUPPORT AXLE

2'-8³/₄"

7¹/₂"   2'-0¹/₄"   2'-0¹/₂"   2'-0¹/₄"   7¹/₂"

6'-1"

Figure 27.3. Chickshaw front elevation.

The hardware-cloth end walls are covered with plastic for extra warmth. (Specific details about temperature and feeding needs for baby poultry can be found in many books and extension pamphlets, or from the poultry supplier.) In colder weather Chickshaws can be started in a garage, then wheeled outside when the weather warms. We get our new chicks every year on June 1.

We open the ramp door to the outside starting the second day to give them access to short grass under a movable protective screened area. We think early access to short grass is the best vitamin pill for growing birds. The chicks can run back inside if they get cold. There are advantages to starting in the Chickshaw. First, the chicks identify with the house and will return to it for shelter after they are let out on pasture. Second, the Chickshaw allows them to have that important early access to grass. The grass area for chicks should be clean ground that has not been grazed by poultry for the past few years. That is one of the benefits of incorporating poultry into a crop-rotation system where clean, new ground can be ensured. We keep moving the Chickshaw and the movable protected screened area across clean sod until the young layers are six weeks old. After that age we let them range uncovered.[2] We add additional Chickshaws to expand their housing as the pullets grow to laying age.

Layers on Range in Chickshaws

The Chickshaws for layers are equipped with a roll-out community laying box and six 2 × 2 roosting bars laid over the mesh floor (so the manure can fall through at night). They are the range shelter from April to November for our mature laying hens. We have four of them for our 200 layers. There is sufficient roosting space in a Chickshaw for 50 mature hens since they are only kept in there at night for protection against predators. Every morning each house is moved its length down the pasture so the manure falls on a new area every night.

When poultry are on pasture, a simple but specific management schedule should be followed: In the morning sprinkle scratch feed outside, open the doors, move the house a short distance, refill the feeders, check on the water supply, and collect any eggs already laid. We use large, covered, outside range feeders and an automatic waterer connected to a hose. The remainder of the eggs are collected in the evening.

The Hoop Coop

During the winter months, November to April, when the Chickshaws would be impractical in this climate, our hens live in a 26-by-50-foot (8 by 15 meter) plastic-covered pipe-frame greenhouse. It sits permanently 2 inches (5 centimeters) above the ground on hard rubber wheels (10-inch/25-centimeter diameter) from Northern Tool, seven wheels on each side. We call it the Hoop Coop. We move it 10 feet (3 meters) every week all winter (October to April) down 320 feet (97.5 meters) of grass/legume pasture to give the chickens 260 square feet (24 square meters) of new sod to scratch in after each move. Before moving we unchain the house; when it is stationary again, it is chained on each corner to a series of T-posts that were driven in solidly alongside the planned path the fall before.

The Hoop Coop has roosts at one end, feeders that hang from the structure, and roll-out laying boxes, which all move with the greenhouse. The waterers sit on a heated base on the ground and are moved by hand. Since there is so much light in a plastic greenhouse (even UV), our hens lay exceptionally well all winter with no need for supplementary lighting.

In April when the Hoop Coop reaches the other end of the field, we transfer all the layers back into Chickshaws for the summer and we use the Hoop Coop as a greenhouse to grow tomatoes, melons, or some other heat-loving summer crop. We till up the strip it has moved down (10 feet [3 meters] per week × 27 weeks = 270 feet [82 meters]) and grow field corn and giant kohlrabi on that well-manured soil. The

Baby chicks started in the Chickshaw out on short grass.

Interior of a Hoop Coop.

kohlrabis, almost the size of soccer balls, are stored in our root cellar; we give a dozen of them to the layers each day all winter, cut in half so the chickens can peck out the flesh. It is their favorite winter food. The field corn is dried and ground for scratch feed.

In October when the summer greenhouse crop is finished we move the Hoop Coop back to the upper end of the field again and align it to travel down the strip next door. In mid-October we transfer our new group of young layers into the Hoop Coop just before they start to lay. The process begins all over again. By late fall, once our new crop of hens are in full production, we sell the old ones as stewing hens or to homeowners as molting layers.

The Hoop Coop offers the following management benefits:

Low initial cost. Since part of the cost of the Hoop Coop is covered by using it as a summer greenhouse, this is the least expensive winter poultry house we can imagine.

Easy mobility. The house is moved effortlessly with a small tractor and a towing yoke. It could also be moved with a hand winch.

Cleanliness. Since the roosts, under which most of the manure is deposited, are in the back 10 feet (3 meters) of the house, it is automatically cleaned each week when we move it. No manure to shovel. At the end of the summer, before it becomes a chicken house again, we pressure-wash the interior. A movable chicken house prevents the buildup of disease and parasites in the environs of the shelter, which never spends too long in one place.

Field corn growing where our winter layers deposited their manure.

CHAPTER TWENTY-EIGHT

The Information Resource

Nature is a language, and every new fact that we learn is a new word; but rightly seen, taken all together, it is not merely a language, but the language put together into a most significant and universal book. I wish to learn the language, not that I may learn a new set of nouns and verbs, but that I may read the great book which is written in that language.

—RALPH WALDO EMERSON

I can see a clear pattern in my development as an organic grower. I have made a conscious shift away from product inputs, which I have to purchase, and have moved toward information inputs that support a farm-generated production system. The more information I acquire about soil improvement, deep-rooting legumes, green manures, crop rotations, and other management practices that correct the cause of soil fertility problems, the less I need to treat the symptoms of low soil fertility by purchasing stimulant fertilizers. Similarly, as I have learned to reinforce plant insusceptibility to pests through improving soil fertility and biological activity,

A few of the books in my farming library.

avoiding mineral imbalance, providing for adequate water drainage and airflow, growing suitable varieties, and avoiding plant stress, I have removed the need to treat symptoms by killing the pests. It is a familiar pattern in my thinking. Information inputs help me focus on cause correction, whereas product inputs only encourage symptom treatment.

As I become increasingly proficient at working with the biology of the natural world, I learn to create crop-friendly ecosystems that mimic natural systems. I like to say that my farming is becoming ever more *biological* (a term I prefer over *organic* to describe this type of farming in general). I have also heard this trend referred to as a shift from shallow to deep organics (to borrow a phrase from the ecology movement). Whatever the name, the understanding is the same. The optimal organic farming system, toward which my farming techniques are progressing, is one that participates as fully as possible in the applicable biological systems of the natural world. Whether my farm-generated-input approach will succeed in attaining its ultimate goal—a time when I need to purchase no inputs at all—is yet to be determined. Given the inherent limitations of the acid podzolic soil with which I began, that goal presents an interesting challenge. But this isn't a religious quest. I don't plan to go out of business by creating some inflexible doctrine for myself. I keep working at it because it's been so successful. With each step in this direction I find myself running ever more stable and productive systems. I like that. However, as I watch organic agriculture begin to move into the mainstream, I notice a strong trend in the opposite direction, toward more, not fewer, purchased inputs. The reasons for this are not hard to find. But first, some definitions.

The Craft, Business, and Science of Agriculture

I define what I do as the *craft* of agriculture. I am a grower. I produce food. Obviously, I prefer to base my production systems on techniques that are both economically and environmentally sound. I prefer information-input (how-to-do) rather than product-input (what-to-buy), not only because the former costs less but also, as I said above, because the results are more successful. But organic growers have to pursue this quest for information-input purely on their own, without outside help or encouragement— and for a very good reason.

That reason is the *business* of agriculture. The business of agriculture is concerned with marketing product-inputs to the farmer. Money is made by convincing farmers they need a product and then selling it to them year after year. The business of agriculture has no interest in telling me about farm-generated systems. E. C. Large in *The Advance of the Fungi* accurately describes the pressures tending to drive farmers toward purchasing product-inputs:

> There was nothing static about the commercial travelers; they pursued the farmers round the dairy, lay in wait for them on market days, bribed them with bread-and-cheese and beer, made demonstrations on their farms, and told the tale about the advantages . . . with an optimistic enthusiasm that made the angels blush for shame. . . . The farmers sometimes paid rather dearly for this form of education . . . the manufacturers were not above charging the farmers as much as possible for some mythical and especial goodness in their particular preparations.

The effective salesman is a force to be reckoned with. But more effective than the sales pitch is the result of using the products. Reliance on purchased inputs encourages growers to ignore natural components of their system that could be nurtured to accomplish the same end. Like addictive drugs, many purchased inputs come close to fulfilling the cynical economist's definition of the industrialist's ideal product, which "costs them a dime, sells for a dollar, wears out quickly, and leaves a habit behind." Economists have told us for years that the increasing cost of inputs is making chemical farming uneconomical and driving small farmers out of business. If that is true, it certainly makes little sense for organic farmers to follow the same trend.

Given the importance I have placed on information-input, you might expect that the third of my categories, the *science* of agriculture, has a lot to offer organic farmers. With the new recognition of organic agriculture by government agriculture departments, useful information should not be far behind. Unfortunately, with a few exceptions, the reality is quite the opposite. The craft of agriculture, that is to say, what each of us does in the field every day, is practical. The science of agriculture, which goes on in laboratories and test plots, is theoretical. As any dictionary will confirm, theoretical and practical are defined as opposites.

In my own search for organic agricultural knowledge, I have become keenly aware of just how wide this gulf between theory and practice can be. Since I enthusiastically endorse the value of information-input, I pay close attention to everything published, read all I can, and keep myself well informed. Yet my conclusion after 50 years of experience is that most of the scientific research being done is not of practical use to me. The explanation is simple. The goals of science and my goals as a farmer are vastly different. Even when we both focus on the same problem, we are not after the same goal. The science of agriculture is concerned with understanding the mechanism; it is problem oriented. The craft of agriculture is concerned with practical application; it is solution oriented. In other words, science asks why, but the farmer asks how. Naturally, different questions lead to different answers. The dean of an agricultural university confirmed this fact recently when he stated that his university was "not involved with how to grow corn but rather with why corn grows."

Now, I am not uninterested in why things work. I use my understanding of the mechanisms every day

to organize my farm operation. But that knowledge does not help me when I need to get my hands dirty. Understanding "why" can help me plan what needs to be done, but it can't help me do it in the most innovative and effective way. At present the science of agriculture is providing us with too much head information and not enough hand information. All of the hand information that I value is coming from other sources.

The theoretical science of agriculture found a role in chemical farming over the past century. Many of the chemical inputs and their development and testing involve the resources of the laboratory—resources unavailable to the farmer. It may also be that the technical complexities of chemical agriculture required the services of scientists as intermediaries to interpret and explain the concepts. But that same situation does not exist with a farming system based on nurturing natural processes. In this case, the people who know best how to do organic farming are those who are doing it every day—the organic farmers themselves. Ironically, their experience is written off as anecdotal evidence and is basically inadmissible to the science of agriculture.

Research of practical use to growers was published prior to 1940. Perhaps in those years the dream of complicated chemical panaceas had not taken hold as thoroughly as it has today. Whatever the reason, many universities back then conducted farm-applicable investigations. I find that material useful but limited. Modern organic growing is not a rehash of old practices. We are moving into new realms. I'm impressed every time I visit Holland, where the agricultural research service continues to produce practical information. Dutch growers have successfully demanded more "use-oriented" data on fine-tuning their cultural techniques. Much of this work is scale-neutral (helpful to small and large growers alike) and system-neutral (helpful to both organic and chemical growers). For example, the Dutch have developed detailed pruning instructions for the fruit trusses of greenhouse beefsteak tomatoes depending on the season of the year.

Since that sort of crop-specific technical knowledge is just what I want for my farm, I have learned to find it on my own.

Finding the Information

I have two main sources of information. First is my own day-to-day experience, which increases in value the more I keep my eyes open and record what I learn. Second is the experience of other growers and agricultural investigators. It is something of a chicken-and-egg relationship, since each source grows out of the other. So let's begin with the second. I have been fortunate to visit fellow growers both in the United States and abroad. I learn a great deal on these visits and owe a real debt to every grower who shares information with me. I have the advantage on these trips of seeing many farms and being able to compare and analyze differences and similarities in

a way that my hosts cannot, because many of them are unacquainted with one another. That overview doubles or triples the value of the individual visits. Whenever you have the opportunity to visit other growers, take it.

Those visits only scratch the surface, though. I couldn't begin to visit everyone, and even if I could, each new idea gleaned from a visit inspires a hunger to expand or refine it. So I do most of my visiting, once a month, at my local state university library. I can gain access to everything I want to know through that service alone. If you are unacquainted with libraries, they may seem as frightening as a large foreign country, speaking an arcane language, with impenetrable local customs. I entreat you to calm your fears. Libraries provide information the way farms provide food. Like farmers, librarians are professionals and proud of their abilities. Whatever information you want to find, all you have to do is ask.

I chose my local state university library, even though it is an hour and a half away, because it has a large collection and, since it is a land-grant institution, that collection includes agriculture. When I lived near the best library in an adjoining state, I used that facility. University libraries are open long hours every day, so I make the most of it by arriving early and staying late. If I expect to need technical help learning a computer search system, or have lots of questions about where to find what I am looking for, I try to plan the trip for times such as school vacations or Saturday football games when most students are elsewhere.

I have nearly always found library staff polite and helpful. Since I am not a student I am usually categorized as a "guest borrower." Depending on the library that may mean I am a second-class citizen and not entitled to all the services of the library. Sometimes the staff does things for me because, since I am older than the students, they assume I must be faculty. I have sometimes worn a coat and tie to reinforce that possibility. Other times, just looking helpless or smiling sweetly can overcome barriers. If I receive a bureaucratic "no" to my request, I will often come back after the shift changes to see if I have better luck with the next person. If all else fails, the little library in my hometown can get me almost any book or article I'm interested in, although it will usually take them a lot longer. So don't let libraries be scary places. They are filled with tons of information and people who can help you locate it.

The resources of the whole world are at your fingertips in the library, now more than ever before. My favorite hunting grounds are the abstract journals such as *Soils and Fertilizers* and *Horticultural Abstracts*. These journals review almost every applicable publication in the world every month and publish short summaries of each article in English along with author, title, and source. Entries are arranged by subject and can be quickly scanned. It may take a visit or two before you determine which subject headings will help narrow your search (there may be more than 1,000 abstracts each month), but you will soon become proficient. You now are in contact with the work of investigators from Egypt to Ukraine to Argentina. The research from some counties, especially the Netherlands and many Eastern European nations, is much more practically oriented than ours.

When you do find a useful article, check out the bibliography at the end. There the author will list previous studies on the subject that were consulted. These will lead you back in time and very often uncover older, more practical investigations or whole new approaches. More than once an older idea the author was discrediting turned out to be more practical and useful to my production system than the supposedly improved model. If, on the other hand, your starting point is an older article, there is a wonderful reference work called *Science Citation Index* that allows you to search for connections toward the present. For example, if your original article appeared in 1972, looking up under the author and title in *Science Citation Index* for any year following 1972 will tell you who listed that article in their bibliography and thus who is investigating that same or a related subject. A day in the library becomes a delightful game of agricultural detective work.

I have pages of handwritten notes from when I first began collecting information years ago. I remember spending long hours searching through library card catalogs. Life is easier now. Photocopy machines allow me to keep the whole article on file at home. Computerized card catalogs make it faster to locate books and also to cross-reference on specific subjects. The Internet gives me access to a world of information and publications without my ever going to the library. My trips now are mainly to read articles in journals to which I have no access from off-site.

There is an overwhelming amount of material in libraries, and it helps to focus your search as much as possible. You will soon learn to skim-read and sniff out the good stuff. The return from all of this is more often inspiration and new possibilities rather than hard facts. But I consider that very worthwhile. On average, 20 percent of what I skim is interesting, 5 percent may be useful, and maybe 1 percent represents real gems—sort of like the one percenters I wrote about in an earlier chapter. It makes the search both rewarding and enjoyable. Each trip to the library results in an incremental improvement in the success of my farm.

My Own Experience

I can remember the fall of 1965. I had decided to begin farming the next summer, and I was reading all the books I could find on the subject. Yet every sentence seemed to create more questions in my mind. I pestered farming friends day and night and was frustrated when I couldn't seem to get my questions answered. It was a typical case of beginneritis. By the end of the next summer, after my first season of actual hands-on experience, the confusion had cleared up. It wasn't that all my questions had been answered. But I now had a basis for knowing what to ask. Many questions disappeared simply because they had answered themselves. Many new ones appeared, and I could not wait to put them to a practical test the next year.

What I learn from experience in the fields on my land is the most valuable input of all. Right from the start I got in the habit of running experiments, and I continue them today. I won't claim they are always well done. I won't claim they are without possible error. It isn't easy to find time for learning when I am flat-out busy. A few times I have had mere seconds to evaluate a test plot because it happened to be growing the only mature carrots or lettuce or whatever we had run out of down at the stand. But just because I cannot publish my ragged results in a professional journal doesn't lessen their value to me. If not useful in themselves, they often provide a hint that leads to some small improvement the following year.

Increasing the Information

So these have been my teachers: the anecdotal experience of other farmers, the day-to-day experience on my own land, and ideas gleaned from books and from my monthly trips to the library. I willingly share what I have learned, because much of it came from others either directly or indirectly. Whenever growers get together and share information, it multiplies exponentially, because each grower brings personal experience from a slightly different perspective. Enormous synergy is generated in the exchange of ideas. By the end no one will remember who suggested what and no one person will get the credit, nor will he or she deserve it. Ideas develop and grow through cross-fertilization. Some other farmer somewhere always has the answer to my question and vice versa. What needs to be done is to move that knowledge from our individual minds to the collective mind.

In the hope of doing just that, a friend and I organized a two-day conference in Vermont a few years ago. We invited the best organic growers from all over the Northeast—as far away as Pennsylvania and New Jersey in the south, New York in the west, and Quebec and New Brunswick in the north. We also invited two exceptional Dutch organic growers. The stated intent of the conference was for the Dutchmen to give detailed presentations on their production practices. The ticket for admission was that each participant had to give a five-minute presentation on the best ideas on his or her farm. We set aside special sessions during the conference for those presentations. It was dynamite. In that room we had the combined knowledge of one heck of an outstanding organic grower.

In selecting the 50 or so participants at that conference, we were careful to choose those who were neither next-door neighbors nor close competitors. We hoped they might be more forthcoming if they felt they weren't giving away secrets to the person next door to them at the farmers market. But I think our precaution was unnecessary. Once we started, not only did everyone realize the benefits of this exchange, but all the participants were forthright because of their justifiable pride in their own accomplishments. I'm sure the same would be true of gatherings in other parts of the country or overseas.

Beginning with my first trip in 1974, I have traveled to Europe seven times to visit organic farms. Over the years I have met and corresponded with exceptional growers from every continent. The information resource that exists on organic farms around the world is truly astounding. If some organization is looking for a program to advance the cause of organic farming, I have a suggestion. Hold a series of meetings like ours in different regions around the world. The meetings would bring together the most skilled organic growers. Instead of asking them what they can't do, ask them what they can do. The idea is to get everyone to share his or her biological or mechanical techniques, solutions, and innovations. The result would be a crop-by-crop instruction manual. I think we would be astounded at how complete that manual is. The only reason this information is not presently accessible to farmers is that it has never been collected and distributed. But the fact is that it does exist.

As organic growers we must recognize the potential of farm-generated replacements for inadequate outside information, just as we have found farm-generated replacements for inadequate chemical inputs. The information for technical improvements in the craft of organic agriculture can be as much a product of the farm itself as is the organic matter for improving soil fertility. The combined efforts of organic farmers can be their own best information service.

Resisting the Future

The history of organic farming is similar to the history of any successful idea that diverges from the orthodoxy of the moment. The orthodoxy first tries to denounce it, then tries to minimize its importance,

A Few Good Information Sources

Growing for Market
PO Box 75
Skowhegan, ME 04976
(800) 307-8949

This is a great newsletter that caters to the needs of the smaller-scale grower. Lots of articles by and about people who are actually doing it.

Alternative Farming Systems Information Center (AFSIC)
National Agricultural Library
10301 Baltimore Avenue
Beltsville, MD 20705
(301) 504-5755

Helps researchers find answers to questions about sustainable agriculture, or obtain copies of an article. Lists of free information products and bibliographies are available by phoning or writing.

Appropriate Technology Transfer for Rural Areas (ATTRA)
PO Box 3838
Butte, MT 59702
(800) 275-6228

ATTRA, a program of the National Center for Appropriate Technology, has a staff of technical and information specialists who will provide customized research or one of about 50 sustainable agriculture information packages, summaries, and resource lists at no charge.

and finally tries to co-opt it. The business and the science of agriculture are like the moneychangers and priests who have lost control of the temple. The organic idea has allowed an increasing number of agricultural heretics to escape from their grasp. Now that organic has become an obvious force, the old order is trying to regain control. But no scientific or business enterprise based on the devising and marketing of miracle products designed to replace nature is going to acknowledge that nature doesn't need to be replaced. I for one do not wish to cede my information gathering to minds still mired in the concepts of industrial agriculture. Nor do I wish to cede my farm's biological future to the wiles of salesmen. The best inputs for organic growing are free and are a function of a whole farm system and its relationship to the surrounding environment.

The leadership of the business and the science sectors of agriculture has led chemical farming down ever-more-tenuous paths. I don't want to see organic farmers sold that same bill of goods. I don't want organic farming to become dependent on its own long list of purchased "natural" inputs, which put the profits in the pockets of middlemen and put farmers on the auction block. It is easy to be co-opted by purchased nostrums, because farm-generated inputs are not running competing ads. The sales pitch for farm-generated inputs has to come from us, the organic growers, because we are the ones who understand the as-yet-untapped potential of biological systems.

Don't forget, the success we have achieved to date is the result of our own efforts. Sure, we have a long way to go, but there are no impossibilities here. It is a mistake to assume limits to organic growing based on what it can or cannot do today. I am using techniques now that I didn't even dream of 20 years ago. I am succeeding at things now that I failed at 10 years ago. I am doing things better now than I did five years ago. These advances are a result of believing that natural biological systems can provide everything my farm needs if I keep exploring them. I didn't go out and blindly impose a vegetable farm on the landscape. I studied, and I continue to study, how to integrate my farm with nature's systems. That integration is an ongoing process.

CHAPTER TWENTY-NINE

The Adventure
of Organic Farming

Seldom in the writings of our highly skilled specialists is there a glimmer of the truth that there is a unity in the health of the soil, the health of plants and animals, and of man. The worship of technology finds little time for a comprehension of nature's laws, or for the humility to understand that we cannot defy nature without repercussions.

—Dr. Walter Yellowlees

For someone like me, whose passions, before I began as an organic farmer, focused on other supposedly impossible activities like rock climbing, mountaineering, and white-water kayaking, organic farming has always felt like an adventure—an adventure into a new part of the natural world—the miraculous part beneath our feet. Exploring the mysteries of the soil doesn't involve high-altitude cold or

We provide fresh local produce year-round.

vertical rock faces or raging rivers, but it still offers the same sense of accomplishment, of satisfaction, and of excitement. So thanks to that adventurer's background, when I first became interested in food and farming over 50 years ago, I was imbued with the adventurer's ethic.

That ethic is crafted on minimalism, respect for the natural world, and independence. Adventurers want to experience the boundaries of the natural world as purely and cleanly as possible, guided by the decisions they make themselves. The ideal in climbing is to avoid all artificiality, to have little need for superfluous technology, and to attain the closest possible intimacy between the adventurers and the reality of the world around them. The dream is to seek out challenges, succeed at doing them, and leave a pristine world for others to follow—to pass through a landscape like sunlight through wind. The goal is in doing it elegantly, and the delight is derived from that accomplishment.

When I was in my 20s, I loved hearing the word *impossible*. I knew that was where I would want to go next. At that stage of my life I supported myself economically with a teaching job, but I supported myself spiritually, on weekends and during school vacations, by adventuring. So when I was introduced to the idea of farming, it seemed very appealing because it would allow me to continue to face challenges in the outdoors. And when I was told simultaneously that farming without chemicals and pesticides was impossible, I was hooked. I wanted to grow food not with artificial industrial aids but in harmony with the same natural systems I had come to know so intimately during my adventures in the wilderness. I was sure that if I investigated natural systems, and understood them, and nurtured them with the adventurer's ethic, the answers would be right there and farming could become an elegant dance, a sensuous partnership between the farmer and the living systems of the earth. And the pure delight I had experienced while engaged in adventures, what the great French climber Lionel Terray referred to as

"sniffing the roses that grow on the borders of the impossible," would now be part and parcel of my daily work. And that is exactly how it has gone.

If that all sounds like fanciful thinking 50 years ago for a 20-something rock climber who had yet to plant a seed, you have to remember that this was the 1960s. The environmental movement was coming into its own. We were enthusiastically idealistic about the planet. Protecting the environment by growing food without chemicals made eminent sense to those of us who loved the natural world. Back in those days the DuPont Company was defending itself with slogans like "Better Living Through Chemistry" because they were aware of the natural human aversion to the image that chemicals represent. Many of us, even though we had no previous agricultural experience, wondered how anyone could do things any more ignorantly, vis-à-vis the planet, than the destructive human activities we were reading about in the papers. *Silent Spring* had made everyone aware of the dangers of pesticides, but even more disturbing than DDT and the legion of apologists from the chemical industry attacking Rachel Carson was the history of pesticide use. Until DDT came along, the leading pesticide was something called lead arsenate, a combination of two very poisonous elements. Maybe a product like DDT might eventually break down, but elements, like lead and arsenic, were in the soil and in the environment forever. To us the idea of an agriculture that had countenanced spreading lead and arsenic on people's food seemed so unconscionable that even as total beginners we were sure we could do better.

Adventurers are very good at solving problems. Whether scoping out a route up a vertical rock face or figuring out how best to pack supplies to camp III, we knew how much planning, foresight, and understanding of the difficulties were required to succeed. We also knew that solutions for one problem had to make practical sense and not have repercussions that led to another. If you pushed too hard, or wanted too much all at once, or were greedy or were in a hurry, you could get into trouble in the mountains. That was obviously true in farming also. But in addition, whereas adventurers have to make smart decisions to keep themselves safe, farming offered a further concern with which we were fully in tune. Farmers have the additional responsibility of making wise decisions that also keep the food eaters safe—the same way my decisions leading a climb had to ensure the safety of my climbing partners. The same unspoken integrity that has always existed among adventurers appeared to parallel all the qualities necessary to become an organic farmer. So I began farming.

Back then we were condemned as Luddites right from the start because we questioned the modern system of agriculture. The scientists and the merchandisers attacked organic farming with a fury. They were outraged that we dared to question their "truth" and threaten their sales. The chemical companies, the US Department of Agriculture, all the land-grant colleges, every extension agent, and even the salesclerk at the local seed and feed store treated us with the type of scorn reserved for the clueless. But we quickly became aware of the Alice in Wonderland world we were entering. Two examples: First, all the old agricultural literature celebrated the indispensable role of organic matter in the soil. Yet we were being told by modern industrial agriculture that soil organic matter was *so* outdated as a driver of soil productivity. By chance a few old brochures made us aware of the existence of a miracle product to save agriculture called "Krilium" developed by Monsanto in the 1950s at a cost of $10 million. Krilium had been marketed as a "soil conditioner." The advertising stressed how its synthetic resins would create crumb structure, aid water infiltration, and improve water-holding capacity. However, as one research paper on Krilium was bold enough to point out, "In effect it functions as a synthetic substitute for the natural gums and resins derived from organic matter in the soil." In other words industrial agriculture knew full well the value of soil organic matter but wanted to sell a synthetic substitute.

Second, at the same time that Firmin Bear of Rutgers was writing that a well-planned crop rotation alone was worth 75 percent of everything else that the farmer did, we were told that monoculture could replace crop rotation. Alice in Wonderland indeed. But as Bob Seger sang, "We were young and strong and we were running against the wind." With that in mind we set out to do the "impossible." And we succeeded right from the start because organic farming has always understood the connection between soil organic matter, soil biological activity, and successful food production.

One of the first lessons we learned was that by tuning in to Mother Nature, we were unstoppable. Since the most of our inputs were delivered by the natural world, our efforts could not be held hostage by some input supplier. The management techniques that maintain natural soil fertility—crop rotation, compost, cover crops, grazing livestock, shallow tillage, and so forth—are information-inputs that enhance natural processes, not product-inputs that have to be purchased. Those techniques create optimal conditions for nourishing the plants and animals that nourish human beings. Since many of us couldn't afford inputs, it was logical for us to be attracted to a system that didn't require them. Self-reliant agriculture, just like the wilderness, attracted self-reliant individuals. But what made it all possible was that we 1960s organic farmers didn't need to prove chemicals wrong. We merely had to prove us right. And we did so with a flourish by growing beautiful and flavorful crops using old-time, natural techniques. Aldous Huxley's statement, "Facts do not cease to exist because they have been ignored," seemed to describe the situation perfectly. The public responded enthusiastically, and the organic market was born.

Fortunately we had great guides as we entered the agricultural wilderness. We didn't have to invent the basic ideas of organic farming. They are a gift from over 100 years of development by wise people who farmed before us. We are the beneficiaries of the intuition, experimentation, and dedicated efforts of

our predecessors who were concerned with the detrimental effects on food quality caused by industrial methods. They developed the art and science of organic farming because they understood that proper nourishment of human beings rests upon the creation of a biologically active fertile soil. Organic farming needs to be defined by the benefits of that fertile soil rather than by its rejection of unnecessary chemicals. Crop resistance to pests and diseases is an outcome of farming a soil that fully nourishes the crops.

Back in 1965 our decision to farm organically was a statement of faith in the wisdom of the natural world and the nutritional superiority of food grown on fertile soils. Somehow we instinctively knew that good farming and exceptional food could only result from the care and nurturing practiced by the good farmer. And the standards we followed, which we learned by talking with and reading the books written by our predecessors, are still my standards today.

First, for uncompromised nutritional value all crops must be grown in fertile soil attached to the earth and nourished by the natural biological activities of that soil. There are so many vital aspects of soil processes that we could not replace even if we wanted to, because we are still unaware of how they all work.

Second, soil fertility should be maintained principally with farm-derived organic matter and mineral particles from ground rock. Why take the chance of bringing in polluted material from industrial sources when fertility can be created and maintained internally?

Third, green manures and cover crops must be included within broadly based crop rotations to enhance biological diversity. The greater the variety of plants and animals on the farm, the more stable the system.

Fourth, a plant-positive rather than a pest-negative philosophy is vital. We focus on correcting the cause of problems by strengthening the plant through optimal growing conditions to prevent pests, rather than merely treating symptoms by trying to kill the pests that prey on weak plants. Extensive scientific evidence is available today on the mechanisms by which a biologically active fertile soil creates induced resistance in the crops.

Fifth, livestock must be raised outdoors on grass-based pasture systems to the fullest extent possible. Farm animals are an integral factor in the symbiosis of soil fertility on the small mixed farm.

The ultimate goal of these five precepts is vigorous, healthy crops and livestock endowed with their inherent powers of vitality and resistance. Add to that the independence that comes from minimal reliance on outside inputs, and you have the adventurer's ethic applied to farming.

These are the standards I absorbed right from the start. These are the standards that were conveyed to the USDA when the National Organic Program was established. However, the USDA quickly showed its true colors by trying to include irradiation, GMOs, and sewage sludge but had to back off because of intense objections from the public. These are the standards that the USDA was fully aware of from their 1980 study, *Report and Recommendations on Organic Agriculture*. That study specified the tenets of organic farming to be as follows: "Soil is the source of life. Soil quality and balance are essential to the long-term future of agriculture. Healthy plants, animals and humans result from balanced, biologically-active soil."

The organic movement began over 100 years ago in the hands of conscientious farmers. No one was forcing them to make the extra effort to be concerned about the nutritional quality of the food they were selling. They did it because it was part of their agricultural ethic. But things are no longer in the farmer's hands. As social critic Eric Hoffer has commented, "Every great cause begins as a movement, becomes a business, and eventually degenerates into a racket." The influence of the farmers who began the movement, the old-time organic growers whose dedication to quality and integrity created the popularity of and belief in the organic label, has been marginalized in favor of expediency ever since "organic" became big business and the marketers and merchandisers took over.

Longtime supporters of organic farming need to realize that the ground has shifted under their feet. Ever since the USDA was given control of the word, the integrity of the USDA CERTIFIED ORGANIC label has been on a predictable descent. The USDA, mired in decades of chemical thinking and influenced by industrial lobbyists, has continually tried to subvert the promise of a natural, biologically based agriculture. The organic community initially insisted on integrity and thought they had achieved it. Unfortunately, the USDA foxes have been managing the henhouse. We now have dairies of thousands of cows and henhouses with hundreds of thousands of chickens with no actual access to grazing and 1,000-acre vegetable fields fed on "soluble organic" fertilizers of suspicious provenance. We also have soil-less hydroponic produce allowed by the USDA to be sold as certified organic with no indication anywhere on the label that it was not grown in the biologically active fertile soil that organic has always required. The only way for customers to be assured of getting real organic food is for them to buy directly from a conscientious neighboring organic farmer.

If you are a certified organic farmer, one simple action you can take right now to reclaim our heritage is to put economic pressure on the certifying agencies to stop certifying substandard crops as organic. Just contact your organic certifier and demand to know if they certify questionable industrial organic operations. If they do certify those sorts of producers, tell them that you object and will plan to change your certification next year to a more principled provider unless they desist. Honestly, why would you want to pay money to an organic certifier whose practices are undermining customer faith in the integrity of the organic label?

If you are an organic eater, speak to the manager in your local supermarket. Say that you object to factory-farm produce and concentrated animal feeding operation (CAFO) meat, milk and eggs masquerading as organic and want the store to stock real organic produce or you will take your business elsewhere. I can guarantee you that most managers will feign ignorance about these issues and tell you that all is well.

Why do I care so much? Why not let the merchandisers run things as they see fit? Hey, it's USDA certified organic, isn't it? Why not be lulled into complacency by the marketing spin doctors in the tasseled loafers who are very skilled at dissembling and fast-talking and injecting the discussion with just enough barely plausible mental Novocain to keep the public quiet. These shysters are no competition for our farm. Our customers love us because we produce the finest food they have ever tasted. So why don't I just stay home and run my farm and forget about the spin doctors?

If you knew what I see on my farm every day, you would understand why I care. I see miracles! I see healthy pest-free plants and animals and clean water and clean air with no need for any chemicals or poisons. I see the impossible being done, because it was never impossible to begin with. I see healthy livestock because they are fed healthy (and appropriate) foods. I see vigorous crops thanks to vigorous soil biology. I see small farmers all over the world with no more resources than the shovel, hoe, and seeds with which I started being taught the simple truths about soil fertility and livestock care, and succeeding. I see well-fed people on a healthy planet.

The successes of organic farming to date are just the beginning of this adventure. What real organic farming will have taught the world about plant health and human health and planetary health in another 10 years will blow your socks off. Research into the soil microbiome is opening whole new vistas for a truly "biological" agriculture. If it is appreciated that when correctly done, the age-old biologically based soil fertility practices that fed humanity for ages (compost, green manures, rock minerals, crop rotations, cover crops, mixed stocking, and so on) result in vigorous crops and livestock that are resistant to pests, a whole new world of agricultural science opens up. The concept of achieving plant and animal health (and, by extension, human health) through enhancing Nature's own systems showcases the potential of the organic farming philosophy to restructure our entire human understanding of the living world. However, that's only if we defend the old-time meaning of the words *organic farming* so there is no diluting of its principles.

What organic farmers have accomplished to date is just the beginning of how the thinking behind organic farming can transform our human relationship with the planet. I want to protect the knowledge gained during the 100-year-long excursion into the organic wilderness, so we can continue the adventure on the right path.

CHAPTER THIRTY

A Final Word

The miraculous succession of modern inventions has so profoundly affected our thinking as well as our everyday life that it is difficult for us to conceive that the ingenuity of men will not be able to solve the final riddle—that of gaining a subsistence from the earth. The grand and ultimate illusion would be that man could provide a substitute for the elemental workings of nature.

—FAIRFIELD OSBORN, *Our Plundered Planet*

There are no simple solutions or shortcuts in this work. No panaceas exist. There are, however, logical answers. Viable production technologies can address environmental and economic realities. Some of these production technologies may require new ways of thinking, while others may appear to revive old-fashioned or outmoded ideas. On closer examination, it will be found that the "outmoded"

Wood for winter heat grows high-quality winter salads.

practices were never discredited, but rather discarded during a period of agricultural illusion when science did seem to promise simple solutions and substitutes.

The production technologies of biological agriculture nurture and enhance the elemental workings of nature. They synthesize a broad range of old and new agronomic practices into an economically viable production system. These technologies are the result of a reasonable and scientifically grounded progressive development, not a return to the old ways. This agricultural system consists of a series of interrelated plant and soil cultural practices that, when done correctly, are no more difficult (albeit obviously more thought provoking) than chemical food-production technologies.

The information presented herein is as up to date as it can be. But it will change. I will modify my approach as I learn new techniques, and I will revise one practice or another. However, I have used these methods long enough to assure readers that they will not go wrong in following my recommendations. Still, each of you will want to change parts of this system. You will want to adapt it not only to fit your own particular conditions, but also to keep from getting into a rut. It is crucial to experiment, adapt, and improve. We owe a great debt to all those farmers and researchers who have gone before us, whose work has either solved problems or left clues that will help us solve them. There are no gurus in this game—no repositories of the "correct way"—only fellow searchers. All the information for further improvement is out there waiting for us to discover it.

The skill that most benefits you as a farmer is learning how to incorporate new knowledge as a productive addition to your present system. Do not hesitate to discard present practices when experience or evidence prove them faulty. But how do you decide? By what criteria can small steps or sweeping changes be judged? In the final analysis, the only truly dependable production technologies are those that are sustainable over the long term. By that definition, they must avoid erosion, pollution, environmental degradation, and resource waste. Any rational food-production system will emphasize the well-being of the soil-air-water biosphere, the creatures that inhabit it, and the human beings who depend upon it.

———◈———

From Artichokes to Zucchini: Notes on Specific Crops

Over the years I have picked up lots of little tidbits of useful information and preferred techniques. For the most part these tidbits are not the standard ABCs (which are well covered by the books listed in the annotated bibliography or by other standard reference works), but rather those little refinements that are the fruits of experience and are so often left untold. The preferences described below for each crop are the best that I have arrived at.

Artichoke, Globe

Vegetatively propagated artichokes are not traditionally grown in cold climates, since the perennial plants won't survive the winter. But artichokes grown from seed offer a commercially viable new specialty crop in those areas. I have grown them in the chilly mountains of Vermont and on the cool coast of Maine. The trick is to turn the artichoke into an annual. All that's required is a little horticultural sleight of hand.

If you plant an artichoke seed, it will usually grow only leaves the first year. The following year it will send up a stalk from which grow the artichokes—which are actually edible flower buds. If the winter is too severe (as winters in most of the northern half of the United States are), the first-year vegetative plants won't survive to become second-year plants. The sleight of hand involves fooling the plants into thinking they are two years old in the first year.

Artichokes.

237

To achieve that, you need to grow the young artichoke plants first in warm, then in cool temperatures. For best results I start the seeds in a warm greenhouse six weeks before the earliest date on which I can safely move them to a cold frame or unheated tunnel. I move them when I am sure the temperature inside the frame will no longer go below 25°F (–4°C). In Vermont I sowed on February 15 and moved the plants to the frame on April 1. I only close the cold frame to protect the plants from hard freezing. The cooler they are for the next six weeks, the better.

The change of growing temperatures from warm to cool (a practice called vernalization in horticulture) is what fools the plants. The first six weeks of warm growing conditions were sufficient time for the plants to complete their first "summer" season. The subsequent six weeks of cool temperatures make them think that they have experienced their first winter. Thus, although they are only 12 weeks old when I transplant them to the field, they think they are beginning their second year. The second year is when artichoke plants begin to produce the flower buds that we eat as artichokes. And so they do.

The care they receive after transplanting will determine the number and size of the artichokes. Under the best conditions, I have averaged eight to nine artichokes of medium to medium-large size per plant. The best conditions are plenty of organic matter (generously mix in compost, manure, or peat moss) and plenty of moisture (mulch thickly with rotted hay or straw, and irrigate regularly).

I space the plants 24 inches (60 centimeters) apart down the center of a 30-inch (75 centimeter) bed. That is much closer spacing than for perennial artichokes, but these plants won't get as large. I harvest by cutting the stem beneath the bud with a sharp knife. Don't wait too long. Once the leaf bracts on the bud begin to open, the flesh gets tougher and more fibrous. In New England the production season is late July through late September—two months of a unique and delicious, fresh and flavorful specialty crop.

Almost any variety of seed-sown artichoke will work to some degree under this system. I have experience with over a dozen cultivars. However, the most successful are those that have been bred for annual production.

Asparagus

This is a perennial crop and therefore is not part of the rotation. A well-cared-for planting can be productive for more than 20 years. I start asparagus from seeds rather than buying roots. Plant early, January 1 to February 1, in order to gain an extra year of growth. Cover the seeds and germinate at 72°F (22°C) in mini-blocks. Pot on immediately to 3-inch (7.5 centimeter) blocks. Grow on at 60°F (16°C) and transplant to the field after the last frost.

Prepare the soil with rock powders and manure. Make sure the pH is up to 7. An extra 50 tons (45,360 kilograms) to the acre (4,000 square meters) of manure, if available, is well spent in preparing for this crop. Set asparagus out in rows 5 feet (1.5 meters) apart, with plants 12 inches (30 centimeters) apart in the row. Make a hole 8 inches (20 centimeters) deep with a posthole digger. Place one soil-block plant in each hole, and fill halfway so the greens are still above the soil. Fill the planting hole to the surface later on, once the greens have grown above ground level.

If the seeds were started early, you can begin a light pick the year after planting. Otherwise, wait until the third season. Cut asparagus spears with a sharp knife just below the soil surface to include a bit of white stem. It keeps longer that way. Cool it immediately after harvest. Store at 32 to 36°F (0–2°C) with 95 percent humidity.

Bean

The difficulty with beans is getting them picked economically. The mechanical bean pickers used on large-scale operations are tough price competition for the small grower who picks by hand. Given that reality, the grower can either treat beans as a loss leader (a crop that needs to be grown to keep customers happy even though it is not economical) or only grow the specialty varieties such as the extra-thin French "filet" types that sell for a sufficiently higher price to justify the picking costs. I recommend the latter, even though these gourmet varieties need to be picked almost every day for highest quality.

Either bush or pole varieties can be grown. The pole varieties may seem easier to pick because they are upright, but a good picker can pick faster with the bush types. Although beans are a legume, they respond well to a fertile soil. Rotted horse manure will grow better beans than any other fertilizer. I grow beans at two rows on a 30-inch (75 centimeter) bed, and I aim for a plant every 6 inches (15 centimeters) in the row. Beans germinate poorly under cool, wet conditions and should not be seeded outside until the soil warms up. For the earliest crop, beans can be transplanted successfully using soil blocks. Use a 1½-inch (4 centimeter) block for single plants and a 2-inch (5 centimeter) block for multiplants (nice for pole beans). Transplant when one to two weeks old.

Beans don't need to be iced after harvest. Wilting can be prevented by high humidity. Store them at 45°F (7°C) and 90 to 95 percent humidity. The containers should be stacked to allow for good air circulation in storage.

Beet

This is a multiseason vegetable. Sales begin with beet greens, then move to baby beets, and on to storage roots in the fall and winter. Different varieties are best suited to different stages. Read the variety descriptions carefully.

I plant storage beets in 16-inch (40 centimeter) rows and aim for a plant every 3 inches (7.5 centimeters) in the row. Early and baby beets are planted in 10-inch (25 centimeter) rows at 2-inch (5 centimeter) spacing. Beet seeds are actually fruits containing one to four seeds, and they need to be thinned. Monogerm

varieties, with only one seed, are available but are not my favorites. The earliest crop can be transplanted in soil blocks. Plant three seeds per block. Thinning isn't necessary, since the dominant seedling in each fruit will usually prevail. Transplant at three weeks. Set out 12 inches (30 centimeters) apart in 10-inch (25 centimeter) rows.

Beets grow best with a neutral pH and an adequate supply of boron. I have found the best answer to supplementing soil boron to be a pelleted product with a 10 percent elemental boron content that I add to the soil before planting. Carefully calculate the amount to be spread. Three pounds of elemental boron to the acre is usually a safe rate for beets in soil with adequate organic matter. Pelleted boron can be spread accurately with a hand-cranked, chest-mounted

Golden beets.

seeder. Beets with greens must be cooled quickly after harvest. Store at 36°F (2°C) with the humidity at 95 percent. Store in well-ventilated containers.

Brassicas

This heading includes broccoli, brussels sprouts, cabbage, and caulifower, because they all have similar growing needs. I find that all four of these crops grow vigorously and are free of root maggot damage if they are grown after a leguminous green manure. In lieu of a green manure, exceptional crops of brassicas can be grown where autumn leaves have been tilled into the soil the previous fall. When the leaves decompose in the spring, they provide a shot of nitrogen for the crop just like the legumes do. In lieu of leaves use alfalfa meal as a fertilizer.

I grow all these brassicas from transplants. These seeds should be covered when seeded in the soil block. They are set out at two rows on a 30-inch (75 centimeter) bed at 16-inch-by-16-inch (40 by 40 centimeter) spacing. Succession plantings will spread the harvest from early summer through late fall.

Broccoli varieties with smaller central heads and better side-shoot production (these are crosses with Asian gailons) have become very popular and are ideal for the salad bar market. Brussels sprout plants can be topped in early fall to encourage even sprout maturity for a once-over harvest. However, for a late fall harvest, choose brussels sprout varieties with good leaf cover to protect the sprouts. Self-blanching cauliflowers make life easier for the grower, but even with them, a leaf should be folded over the head to provide more ensured blanching conditions. Long-standing cabbage varieties give the grower some leeway in scheduling cabbage harvest.

Harvest broccoli, cabbage, and cauliflower by cutting the stem with a sharp knife. Snap brussels

Broccoli.

sprouts off the stem with a quick side motion. To harvest sprouts for storage, remove the leaves from the plant, cut the whole stem, and store with the sprouts attached. We also sell them like that at our fall and early-winter farmers market, and they are very popular. Broccoli, brussels sprouts, and cauliflower should be cooled quickly after harvest. All the brassicas keep best stored at 32°F (0°C) with a humidity of 90 to 95 percent. Excellent aeration in storage is important.

Carrot

I have noticed over the years that consumers can readily distinguish the superior flavor of organically grown carrots more than any other vegetable. That is not surprising, since petroleum distillates are used as herbicides on conventional carrot crops, and they taste like it. Furthermore, studies have shown that carrots

Tender carrots.

take up pesticide residues from the soil and concentrate them in their tissues. The quality grower can truly excel with this crop by growing succession plantings and selling fresh over as long a season as possible.

Varieties should be chosen for flavor. We plant our earliest crop in January as soon as a bed is available in an unheated tunnel, such as following leeks. A late-planted crop can be covered with an unheated tunnel and baby carrots harvested from under a floating row cover throughout the winter. These will be the sweetest, tenderest carrots anyone has ever eaten.

Carrots for fresh sale are planted at 12 rows to a 30-inch (75 centimeter) bed with either the four-row or six-row seeders. We use that same spacing for greenhouse and outdoor carrots. I sow storage carrots at four rows to the bed and aim for 1-inch (2.5 centimeter) spacing. Either pelleted or naked seed can be used, depending on the seeder. It is worth running your own germination test on pelleted seed before using. Germination can be disappointing. For dependable germination of carrots, it is vital to keep the soil moist from the time of seeding until they emerge. I always direct-sow carrots because I have never found a dependable system for transplanting this taprooted crop.

Don't plant carrots in a weed-infested soil. The in-row weed problems will be overwhelming. On a reasonably clean soil, we use a stale seedbed and we flame six days after sowing. In the fall it is a good idea to hoe soil up over the shoulders of mature storage carrots to forestall greening and as extra protection against freezing prior to harvest. Carrots can be loosened in the soil by using the broadfork to make them easier to pull. When we bunch carrots, we cut off the upper half of the tops after tying to prevent excess moisture loss. Bunched carrots must be kept moist and cool to keep the roots from wilting. We sell our winter baby carrots loose with 1½ inches (4 centimeters) of green top left to show they are freshly harvested.

Store carrots at 32°F (0°C) with a relative humidity of 95 percent. Well-grown, mature carrots will keep in excellent condition for six months. Stored carrots can turn bitter when stored with apples, pears, tomatoes, melons, or other fruits and vegetables that give off ethylene gas in storage.

Celery

Celery is not a common crop on the small farm, but I believe it should be. There is a ready market for organically grown celery. The ideal conditions are a highly fertile soil with lots of organic matter and a steady supply of moisture. The grower must make sure that the soil contains adequate supplies of calcium and boron. If you are near a hatchery that has egg wastes, they are a great soil improver for celery. Irrigation is a must for successful celery crops.

Start celery in mini-blocks at 72°F (22°C). Mist the blocks frequently until germination occurs. Pot on to 2-inch (5 centimeter) blocks and grow at 60°F (16°C)

Celeriac.

or warmer. Sustained temperatures below 55°F (13°C) after transplanting to the field will cause celery seedlings to bolt to seed in the first year. Transplant celery into 12-inch (30 centimeter) rows at 12-inch spacing in the row. Early outdoor transplantings must be protected with a tunnel to keep them from bolting to seed if the weather turns damp and cool. I grow my earliest crop along the edges of the early-tomato greenhouse where the roof is not tall enough for staking tomatoes.

Store celery at 32°F (0°C) with a relative humidity of 95 percent. We use a small mister to maintain storage room humidity. The same advice applies to celeriac. Celeriac, however, is slightly more forgiving about bolting and moisture requirements.

Chard

Swiss chard is a relative of the beet and responds to similar growing conditions. I start chard in 1½-inch (4 centimeter) blocks and thin to one seedling per block. I transplant at three rows to a 30-inch (75 centimeter) bed at 12 inches (30 centimeters) in the row for a cut-and-come-again harvest, or into 12-inch (30 centimeter) rows at 6-inch (15 centimeter) spacing for a onetime harvest of the whole plant. I think the eating quality is better from the latter, but a market that demands large leaves with prominent midribs will require the former. Like most leafy greens, chard does not store well. Cool it quickly after harvest, and keep the humidity as high as possible.

Corn

This crop causes a frustrating decision for the small grower. Financial return per acre from sweet corn is low, but the popularity and demand for the crop are high. What to do? If you are marketing to subscribers, you will want to grow just enough to meet your responsibilities. If you are growing a more limited list of crops, corn is not likely to be one of them unless you have extra land. If you market at a stand, sweet corn is one of the crops you can buy in from other growers.

Corn can be transplanted for the earliest crop by using soil blocks. Place the blocks in a bread tray. With the mesh bottom, air gets to all sides of the block and prevents the taproot from growing out of the block. Do not plan to hold corn seedlings past seeing the first shoot emerging from the top of the block before transplanting. Outdoor plantings are dependent on the temperature. Corn germinates poorly if the soil temperature is under 55°F (13°C). Some growers are experimenting with pre-germinating the crop before planting in order to make early seeding more dependable. All you want is to break the seed dormancy, not have a root sticking out. The idea looks promising, and I suggest that other growers may want to try their hand at it.

For a continuous supply of sweet corn, the grower should plant a number of varieties with successive harvest dates. Some experience will be necessary to determine the best varieties, since in practice varieties do not always mature as progressively as the catalog information indicates. The sugar content in corn begins to decrease after harvest, so the fresher the corn, the better the flavor. If you wish to preserve that sweet corn quality, it is most important to cool the crop quickly and keep it cool. The new super-sweet corns do not lose their sweetness as quickly after harvest, but I think there is much more to the flavor of corn than just sweetness. I do not believe these varieties are as nutritious, either.

Cucumber

This is a warm-weather crop, one that grows best in the most fertile soil you can provide. Greenhouse cucumber growers have always used more manure and compost for this crop than for any other.

Sweet corn grown in hills.

'Socrates' cucumbers.

Composted sheep and horse manure have been the favored soil amendments. Cucumber pest problems are usually a result of imbalanced fertilization—excess nitrogen from chicken manure or a lack of trace elements. Most cucumber-growing problems can be cured by amending the soil with a well-finished compost plus dried seaweed for trace elements. The results are worth the effort.

Many home gardeners grow cucumbers. Your market will be determined by surpassing their quality. This is best done by growing one of the European-style, thin-skinned varieties. Most of these need to be grown in a greenhouse or tunnel. They must also be trellised and pruned to one stem and to one fruit at each leaf node.

Growing cucumbers vertically pays off in yield per square foot. Trellising the crop upward makes the most efficient use of your highly fertile and best-protected growing areas. The plants can be trained up a length of strong garden twine. At the top of the support, the plants are pruned to two stems that are trained over the support. The vines then descend back to the ground while continuing to produce cucumbers. Total production of the greenhouse varieties can reach 50 cukes per plant.

I start all cucumbers as transplants. Germinate them in 3-inch (7.5 centimeter) blocks at 85°F (30°C) bottom heat and grow at 65°F (18°C) air temperature. Drop the bottom heat temperature to 70°F (21°C) after germination. Transplant them to the greenhouse or tunnel at two to three weeks of age. In the greenhouse or tunnel, I plant one row of staking cucumbers in a 30-inch (75 centimeter) bed at 24-inch (60 centimeter) spacing in the row. Give them plenty of water and plan to top-dress with extra compost at monthly intervals. Pick them every day to keep the quality high. Overgrown cucumbers put an extra strain on the plants and lower the yield. Store cukes at 50°F (10°C) with a relative humidity of 90 to 95 percent. Ethylene from apples, tomatoes, and other produce will cause accelerated ripening and cause the green color to change to yellow.

Garlic

Fall-planted varieties are a better bet than spring-planted. The fall varieties are planted in mid-October, winter over in the soil, and mature in summer. There is time for a green manure or succession crop to be established after harvest. My experience with garlic varieties is that they can be very specific to soil types. When you start out growing garlic, you should try as many different cultivars as are available and then choose the one that does the best on your soil. After that, save your own planting stock every year and select for large-sized bulbs. Garlic must be well cured and dried after harvest. It stores best at 32°F (0°C) and at a humidity of 65 percent.

Kale

Kale is a relative of the cabbage family, and the same soil fertility suggestions apply. I grow all kale as transplants in 1½-inch (4 centimeter) soil blocks and set them out at three rows on a 30-inch (75 centimeter) bed at 12-inch (30 centimeter) spacing.

Planting garlic.

I have sold kale both as bunches of leaves and as whole plants. Kale is most flavorful in the fall, after a few light frosts, so I plant it as a succession crop. Since brassicas grow very well following a member of the onion family, fall kale could be the ideal crop to follow garlic. Kale can be left in the ground and harvested right up through hard frosts. Any kale still around very late in the fall makes a tasty green treat for laying hens even when frozen.

Lettuce

Lettuce, in contrast with sweet corn, has a very high dollar return per square foot of crop. On some intensive market gardens, lettuce is the major crop. That complicates crop-rotation planning. Extra compost is needed where the same crop is grown at too short an interval.

I grow all lettuce from transplants for a number of reasons. First, I want to be sure of a full crop without gaps in the rows. Since lettuce seed germinates poorly in hot, dry weather, I prefer to sow it under controlled conditions indoors. In very warm weather

'Winterbor' kale.

the seeded blocks can be germinated in two days in a cool cellar and then brought up to grow. Second, since most lettuces are a 60-day seed-to-harvest crop, they can spend a third to a half of that time as seedlings before transplanting. During the three to four weeks the lettuce seedlings are in the blocks, an unrelated crop can occupy that same ground. This not only increases production but also lessens crop-rotation problems. And third, lettuce is a fast-growing crop, and I want a vigorous seedling grown under ideal conditions to go into a fertile soil and grow quickly. The excellent lettuce transplants from soil blocks give me just that. Information on the timing of succession lettuce plantings is presented in chapters 8 and 26.

I grow lettuce for heads at a 10-by-10-inch (25 by 25 centimeter) spacing in the greenhouse and outdoors at 12 by 12 inches (30 by 30 centimeters). The multileaf 'Salanova'-type varieties are transplanted four rows to a 30-inch (75 centimeter) bed at 8 inches (20 centimeters) in the row. The outdoor lettuce receives an application of well-finished compost lightly mixed into the topsoil just before transplanting. The key to successful lettuce culture is quick growth. You want to make all the growing conditions as ideal as possible. That means not only soil preparation and irrigation, but also high-quality transplants. Treat the lettuce transplants gently so they go into the ground without torn leaves or soil-filled hearts. When the weather conditions are too warm, you will want to harvest the lettuce at a younger age to keep up the quality and double-bag them (sell two smaller heads for the price of one).

Melon

Melons are another crop that requires warm weather, even more so than cucumbers. Any extra heat that can be provided through soil-warming mulches, row covers, windbreaks, or a sheltered site will pay off in a better outdoor crop. Row covers over Quick Hoops make for excellent melon growing. The cover should

'Gold Star' melon.

Onions sown at four seeds per block.

be removed at pollination time. The same soil conditions apply here as for cucumbers. The more fertile the soil, the better the melon crop. A sandy soil is often preferred for melon growing, since it warms up more quickly. Extra organic matter helps a sandy soil hold water.

Muskmelons should be harvested at full slip (the stem separates easily from the fruit) for best flavor. Many of the European Charentais varieties have to be harvested before slip to be at their best. Be sure to read the catalog descriptions of leaf color changes. Melons can be stored at 40°F (4°C) at high humidity for a short time after harvest.

Onion

This is the crop that taught me the value of multiplant blocks. Since I prefer a round globe-shaped onion (as opposed to the fattened globe from sets), I grow onions from transplants. But transplanting and subsequent in-row weed control were not as efficient as I would have liked. Multiplant blocks changed all that.

With multiplant blocks four or five seeds are sown in each 1½-inch (4 centimeter) block five to six weeks before the earliest outdoor transplanting date. Onions, like the brassicas, grow better seedlings if the seeds are lightly covered with potting soil. The blocks are set out at three rows on a 30-inch (75 centimeter) bed at 12 inches (30 centimeters) in the row with the same equipment used for other block transplants at that spacing. In-row weeds are no problem, since the space between the blocks can be cultivated in both directions. The onions grow together in the clump, pushing one another aside for more room as they get bigger. At harvest, each clump is a circle of four to five onions the same size as if they had been spaced normally in the row.

Onion-family plants are greatly affected by the preceding crop in the rotation. The most favorable preceding crops are a fine grass (red-top), lettuce, or a member of the squash family.

I use the same multiplant technique described in the chapter 17 for scallions, but at an even higher density. Twelve seeds are sown per 1½-inch (4 centimetes) block and set out at four rows to a 30-inch (75 centimeter) bed at 8 inches (20 centimeters) in the row. At maturity, the clumps are already prebunched for harvest.

The onion crop is ready for harvest when the leaves begin to die down naturally. If weather conditions are good (warm and dry), we pull the onions, cut the tops, and leave them to dry in the field for a few days. We then cure the onions in a single layer on the benches at one end of our seed-starting greenhouse with fans blowing the air around. Heat can be provided by the plant-house heating system to dry the air if conditions are too moist. The complete drying procedure will take three to four weeks or so, until the necks are completely dry. After drying, store at 32°F (0°C) with a relative humidity of 65 percent. These practices will give you the highest-quality onions for sale and storage.

Parsnips.

with cool temperatures and high humidity for a month or so, but I prefer to harvest and sell it fresh.

Parsley

This is one of my favorite foods, and I snack on it while I work. I think that with some marketing effort, it could be a much more important crop for the small grower than it is. The choice of flat- or curled-leaf varieties will be determined by the market.

I grow all parsley from transplants. It is started in mini-blocks at 72°F (22°C) and then transplanted on to 2-inch (5 centimeter) blocks. Soil blocks are the only consistently efficient and dependable way to transplant a taprooted crop like parsley. I set out succession crops of parsley in any odd corner, both in the field and in the greenhouse. I harvest by crew-cutting the whole plant and then letting it regrow new greens before the next cut. Parsley can be stored

Parsnip

Even though I enjoy eating parsnips and grow them for my own table, I do not think that the return from this crop is sufficient to justify including it in any but a guild marketing system. Parsnips need to be planted early, cultivated through the season, and ideally left in the ground to be harvested early the following spring. The cold winter temperatures turn some of the starch to sugar and make parsnips a real spring treat for those who enjoy them.

An inexpensive precision seeder like the Earthway can plant raw parsnip seed adequately, although the rows will need to be thinned. For other seeders, like the Jang, pelleted parsnip seed is recommended. I plant parsnips at two rows on a 30-inch (75 centimeter)

bed and aim for a spacing of 3 inches (7.5 centimeters) in the row. If a market exists, parsnips do have the advantage that they can be harvested and sold at a time of year when there is traditionally very little farm income. It is important to dig them just as soon as conditions permit and before they begin to sprout. Fall-harvested parsnips that can be held at a temperature of 32°F (0°C) for two weeks in storage can attain a sweetness close to those left over the winter in the field. The humidity level must be kept at 95 percent.

Pea

Peas have more variability than almost any other crop. There are low varieties, tall staking varieties, smooth-seeded, wrinkled-seeded, regular peas, snow peas, and sugar snaps. The problem with peas, as with beans, is getting them picked economically. For

that reason, the more exotic types like sugar snaps and snow peas may be the best bet in some markets, because their prices better reflect the real costs.

I grow both the early low-growing varieties and the later staking types in order to spread out the harvest season. I plant all peas in double rows 6 inches (15 centimeters) apart down the center of a 30-inch (75 centimeter) bed. With the low growers, the two rows lean against each other to keep them more upright. With the staking peas, the rows climb either side of a 6-inch mesh netting supported by EMT posts and crossbars.

Peas can be transplanted for an extra-early crop. Plant four seeds in each 2-inch (5 centimeter) block. Place the blocks in bread trays with mesh bottoms to get air to all sides and prevent root emergence. Transplant as soon as possible.

Peas need to be harvested frequently (at least every other day) for highest quality. They should be

Trellised peas.

rapidly cooled to 32°F (0°C) right after picking. I suggest selling them the day they are picked. Old peas are bad business.

Pepper

This is another warm-weather crop that will repay you for any climate improvement you can provide. Floating covers, plastic mulch, and field tunnels will all aid the production of the pepper crop. I have used the spaces between greenhouses to provide a warm microclimate for a rotation of peppers, melons, and celery.

I start peppers in mini-blocks at 72°F (22°C). They are potted on to 2-inch (5 centimeter) blocks and then to 6-inch (15 centimeter) pots in order to grow the finest early transplants. Nighttime temperature minimum is 62°F (17°C). I do not let fruit set on the plants before I transplant the blocks to the soil. I also pinch out the lowest pepper blossom. I get much greater production later on by reducing that early strain on the plants. It is best to avoid highly nitrogenous soil amendments like chicken manure. The extra nitrogen makes the pepper plants go more to leaf than to fruit.

After harvest, peppers should be stored at 50°F (10°C) at a humidity of 90 to 95 percent. Temperatures below 45°F (7°C) predispose these hot-weather fruits to bacterial decay.

Potato

The best return in potato growing comes not from the main crop but from extra-early harvest of baby new potatoes. That is especially true for one of the yellow-fleshed varieties like 'Charlotte,' 'German Fingerling,' or 'Rose Gold.' The gourmet market will pay handsomely for the crop, and the field is then

'Red Ace' pepper.

'Charlotte' potatoes.

made available for a succession planting of another vegetable or a green manure.

I plant potatoes down the center of a 30-inch (75 centimeter) bed at a spacing of 8 to 12 inches (20 to 30 centimeters), depending on the variety and the size desired. I pay a great deal of attention to the rotational position and soil fertility for potatoes. I do not grow them at a low pH, but I try to prevent scab by providing excellent potato-growing conditions and preceding them in the rotation with a scab-suppressing crop. The most destructive pest in my part of the world is the potato beetle (*Leptinotarsa decemlineata*). Until 1987 it was the one pest problem I had not figured out. In that year we began specific trials to determine the stresses on potatoes and what cultural practices would help us avoid them. We found that mulching heavily with straw just after potato emergence reduced the potato beetle problem by 90 percent or more. It seems that too warm a soil and a fluctuating moisture supply are major stresses on potatoes, and both were minimized by mulching. Varied bits and pieces of published research are in agreement with that conclusion. Chisel-plowing the potato field to break up any hardpan and improve soil aeration for deep root growth has also had a positive effect. I am continuing to work on perfecting the system and also on growing the mulch as a thick cover crop the fall before and planting the potatoes through the residues.

A one-row potato digger can be purchased as an attachment for walking tractors. It is a worthwhile investment if you grow many potatoes. For best storage, potatoes need a period of two weeks at 50°F (10°C) to heal cuts and bruises. After that, storage temperatures of 40 to 45°F (4–7°C) with 90 percent humidity will keep them in fine shape. Storage temperatures below 38°F (3°C) tend to make the potatoes undesirably sweet through a change of some of the starch to sugar. Potatoes that have been stored in too cool a place can be reconditioned by holding them at 70°F (21°C) for about two weeks before use.

Pumpkin and Squash

I include pumpkins and squashes together, since their growing requirements are similar. Both crops thrive on a fertile soil with lots of organic matter. Both are vining crops, and both are planted in widely spaced rows. Pumpkins may only be valuable as a Halloween crop. I think the best "pumpkin" pies are actually made with winter squashes.

These are good crops to plant on weed-infested land. Since they are frost-tender and thus planted late, the field can be cultivated a couple of times before planting to initiate the weed-seed germination and control process. After planting, the wide spaces between the rows and the slow early growth of the plants before they begin to vine provide further opportunities for clean cultivation. Finally, once the vines and large leaves begin to cover the ground, they do a pretty good weed-smothering job of their own.

We start both pumpkins and squash at two seeds per block in 3-inch (7.5 centimeter) blocks and transplant them in two to three weeks. In some short-season areas, starting the plants ahead may be the only way to ensure full maturity of the fruits. I plant in rows 8 feet (2.4 meters) apart and aim to have a vigorous pair of plants every 6 feet (1.8 meters) in the row. Transplanting stresses squash-family plants, and they need protection from cucumber beetles for the first few weeks in the soil (see chapter 21).

It is best to harvest just before frost in fall. Frost damage can inhibit the keeping qualities of the fruits. Carefully cut them from the vines to leave a sturdy stem. After harvest, we cure our winter squashes for a couple of weeks or so at 80°F (27°C) in our seed-starting greenhouse. Store at a temperature of 55°F (13°C) with a relative humidity of 65 percent. Hubbard squashes are less liable to storage rot if the stems are completely removed before storage.

Radish

A well-grown radish is a wonderful salad vegetable. In order to fulfill its potential, it must be grown quickly in a very fertile soil. Growing temperatures on the cool side will help. I like to grow radishes as a late fall and early spring crop in unheated field tunnels or the mobile greenhouse.

We plant radishes at 12 rows to the bed with the four-row or six-row seeders. Succession sowings every week ensure the best quality. If root maggots have been a problem, use the same autumn leaf fertilizer as for brassicas. Till leaves under in fall (up to 100 cubic yards to the acre), and till again before planting. Leaves used as a soil amendment this way release nitrogen in a form that seems to help the radishes grow right past their pests. Alfalfa meal, spread as a fertilizer for radish crops, has also given extraordinary results.

Radishes should be cooled quickly after harvest. Don't bunch them with the tops on except for immediate sale. The tops will expire moisture and cause the roots to wilt. The crisper the radish, the better the sale.

Spinach

The trick with spinach is not growing it in season, which should be relatively easy on a fertile soil, but growing it out of season. Spinach is a cool-weather crop, but the demand for it as a salad component and as an ingredient in many gourmet dishes extends year-round. There are a number of ways to meet that demand.

A clay soil has more body and is a better choice for growing hot-weather spinach than a sandy soil. A sandy soil can be improved with plenty of compost, additions of clay, and a good irrigation system.

French Breakfast radish.

Spinach transplants for an extra-early crop.

'Big Beef' tomatoes.

Under difficult conditions, spinach responds to the same feeds as celery, so egg wastes or crab shells are also effective as a fertilizer. If a spinach crop is planted toward the middle of September and then protected with an unheated movable greenhouse, the harvest can extend through the winter months.

Spinach is not often transplanted, but it is easy enough to do. Sow four seeds per 1½-inch (4 centimeter) block. Transplant three-week-old seedlings every 6 inches (15 centimeters) in rows spaced at 12 inches (30 centimeters). Instead of being transplanted, however, the early crop is usually sown the fall before and wintered over. Young spinach plants are quite hardy and will normally survive the winter with no protection if there is a covering of snow. If snow isn't dependable, a light mulch of pine branches or a floating cover will provide the extra protection. The spring crop can be speeded up by placing a field tunnel over it as soon as the ground thaws.

We harvest overwinter spinach by cutting leaves with a sharp knife. Regrowth for subsequent harvests is better this way than by crew-cutting the whole plant. I saw a very efficient system for outdoor spinach harvest in Holland a number of years ago. The spinach was planted in close rows, and it grew very thick and upright. Harvesting was done by mowing the whole bed carefully with a scythe equipped with a net cradle to collect the spinach.

Spinach should be cooled quickly after harvest and covered with crushed ice if possible. The relative humidity should be in the 90 to 95 percent range.

Tomato

When a selected tomato variety is well grown and fully ripened on the vine, there is no more appealing snack. Vine-ripened greenhouse tomatoes, along with lettuce, are the two most remunerative crops for the intensive salad grower. I most emphatically do not mean "greenhouse tomatoes" in the sense that consumers have come to regard them. These are not the tasteless, plastic-looking objects with no flavor that were picked green, ripened artificially, and sold for looks alone. I'm talking about real tomatoes grown in a fertile greenhouse soil in order to extend the season and improve quality.

I recommend growing greenhouse tomatoes for a number of reasons. The tomato is a popular crop, but the outdoor season is short. Greenhouse production can greatly extend that season and bring customers to your farm. Many of the diseases of outdoor tomatoes, such as blight, are related to weather stress. Under both greenhouse and tunnel production, those stresses are lessened or non-existent, and blight is not a problem. Much of the eating quality of a warm-season crop comes from ideal weather conditions. In northern climates, those conditions do not usually exist. In southern climates they may not exist long enough to fully extend the tomato season. In most cases, long-season production of vine-ripened tomatoes under controlled conditions is a viable option.

I grow mainly beefsteak varieties. Seeds can be planted as early as mid-October on if you are specializing in greenhouse tomatoes. I sow my first crop in mid-February. Seeds are started in mini-blocks at 72°F (22°C). After 8 to 10 days they are potted on to 2-inch (5 centimeter) blocks and grown at 62°F (17°C) night temperature. After another 10 to 14 days they are potted on once more, this time to 6-inch (15 centimeter) pots. I transplant to a heated greenhouse when the plants are six weeks old.

I grow greenhouse tomatoes down the center of 30-inch (75 centimeter) beds that are heavily amended with compost. I grow at a night temperature of 62°F (17°C), and I ventilate at 75°F (24°C). The soil temperature should be no lower than 60°F (16°C) for best growth. If you don't have soil heating, you can warm the soil by covering the beds with clear plastic for at least two weeks before planting, and leave it on until the first compost top-dressing about six weeks later.

I grow tomatoes at a 2-foot (60 centimeter) spacing in a single row down the center of the bed. Prune to a single stem, and remove side shoots every few days. Trellis the plants to overhead supports; tie strong twine (untreated) loosely to the base of the plant and attach it to the overhead support. As the plant grows, the twine can be twisted around the stem, or the stem can be supported with special clips. According to Dutch research on beefsteaks, for optimal production without stressing the plants, the fruit clusters should be pruned to three fruits on the first two clusters and four fruits thereafter for spring crops (three fruits on the first three trusses, two on the fourth and fifth, and three fruits on the sixth and thereafter for fall crops). A top-dressing of another inch (2.5 centimeters) or so of compost should be added to the soil every six weeks. Irrigate regularly.

When the plants reach the overhead support (ideally, 8 feet/2.4 meters above the ground), either they can be stopped for a short production season or the plants can be lowered. In order to lower the plants (a practice known as lowering and leaning), you will want to have left an extra length of twine at the top. Greenhouse-supply companies sell special hanging bobbins for this purpose. By this stage of growth, the bottom cluster of fruit will have been harvested, and the lower stem will be bare. Loosen the twine at the top and let out some length to lower the plant so the bare stem approaches the soil. Move the twine attachment point down the row toward the next plant. Do this to all the plants in the same direction along the row. At the end of the row, start training the plants around the corner and then back down the other side of the bed. Each time the plant top reaches the support, it should be lowered in this same way. As long as the top 7 feet (2 meters) or so of the plant is vertical, it will grow normally. The bare plant stems will eventually contact the soil and send out new roots where they touch. (If you are using grafted plants, you need to keep the stems elevated above the soil so they do not root.) An early planting can be kept productive right through to late fall by lowering and leaning. For more detailed information consult the books on greenhouse tomato growing in the bibliography and contact other, more experienced growers. The specialized techniques for this crop are advancing very quickly.

The first harvest (from a February 15 sowing) is around June 1, and we continue until the middle of November. I pick these tomatoes only when they are vine-ripe and ready to eat. The favor will bring customers back in a steady stream. After mid-November there is not enough sun, and the taste declines precipitously, so we pull the plants and shift to growing short-day crops in the winter greenhouse

Witloof / Belgian Endive

Belgian endive (witloof) is a member of the chicory family, which includes radicchio, escarole, frisée, and curly endive. It has a crisp texture and a sweet, nutty flavor with a pleasantly mild bitterness—great served raw or cooked.

Witloof is one of the most interesting vegetables to grow. It requires a two-step process before it is ready to eat—growing the root and then forcing the root. The first growth takes about 150 days in the field where the chicory grows from seed into a leafy green plant with a deep taproot. At harvest the leafy tops are cut off. We pull the roots, cut them to 7 inches (18 centimeters) in length, and place them side by side, upright, in 8-inch- (20 centimeter) tall bulb crates. We stack the crates in our cold cellar, where the roots enter a dormancy period. For the forcing process we have a number of opaque plastic tubs into which we place a root-filled bulb crate. We add 4 inches (10 centimeters) of water in the bottom and place the tubs on a heat pad at a temperature of 65°F/18°C (fall) and 55°F/13°C (spring). We cover them carefully with heavy-duty black plastic bags to exclude all light while the witloof shoots are growing out of the top of the root. This blanching prevents the extreme bitterness associated with eating green chicory leaves. The witloof "chicons," as

they are called, are ready in 14 to 28 days depending on variety, temperature, and time of the year. They are a consistently popular product at our winter farmers market.

Zucchini

I will include yellow summer squash in this category as well. These are crops whose extra virtues are beginning to be appreciated. There is an eager market for fruits picked small (à la the French courgette). There is also a market for fruits picked with the blossom attached and for male blossoms. Special varieties have been developed for the blossom market.

As with other fast-growing crops, these squashes will thrive under the best growing conditions you can provide. They can be transplanted to field tunnels for an early crop. For the outdoor crop, I prefer direct-seeding. The main key with zucchini and summer squash is to pick them on time. If you're picking small fruits, pay close attention to the plants. These squashes are only valuable when young and fresh. They are so productive that a new harvest is ready every day, and even twice a day in hot weather. Handle the fruits very carefully when harvesting. Some growers wear soft gloves to avoid scratching the tender and easily bruised skin. Don't plan to hold them for long after harvest.

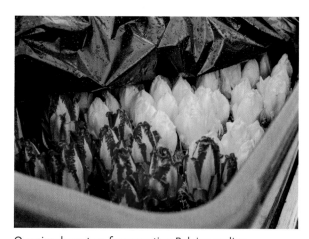

Our simple system for sprouting Belgian endive.

USDA North American Hardiness Zone Map

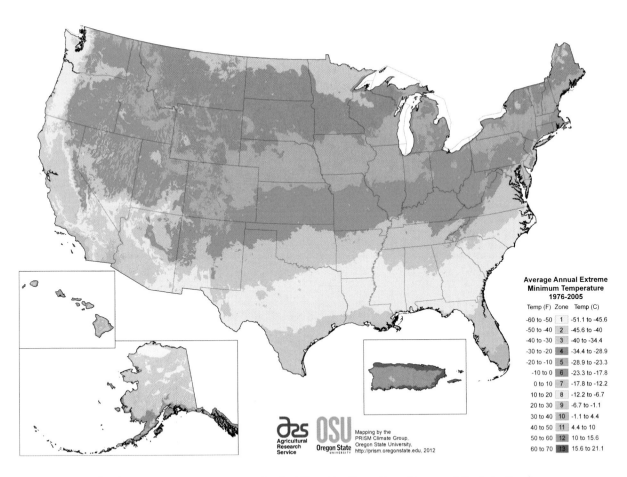

Average Annual Extreme
Minimum Temperature
1976-2005

Temp (F)	Zone	Temp (C)
-60 to -50	1	-51.1 to -45.6
-50 to -40	2	-45.6 to -40
-40 to -30	3	-40 to -34.4
-30 to -20	4	-34.4 to -28.9
-20 to -10	5	-28.9 to -23.3
-10 to 0	6	-23.3 to -17.8
0 to 10	7	-17.8 to -12.2
10 to 20	8	-12.2 to -6.7
20 to 30	9	-6.7 to -1.1
30 to 40	10	-1.1 to 4.4
40 to 50	11	4.4 to 10
50 to 60	12	10 to 15.6
60 to 70	13	15.6 to 21.1

Agricultural Research Service

OSU Oregon State UNIVERSITY

Mapping by the
PRISM Climate Group,
Oregon State University,
http://prism.oregonstate.edu, 2012

The 2012 USDA hardiness zone map. For a more detailed Plant Hardiness Zone Map, see http://planthardiness
.ars.usda.gov/PHZMWeb/Maps.aspx.

—◆◆◆—

A Schematic Outline of Biological Agriculture

I originally prepared this outline as an exercise to compose my thoughts and see if everything fit together. Did it make sense on paper as well as in the field? An early grade-school teacher once told me that if the thinking behind my idea was fuzzy, I wouldn't be able to make a coherent outline.

This is pretty coherent. She would have been pleased.

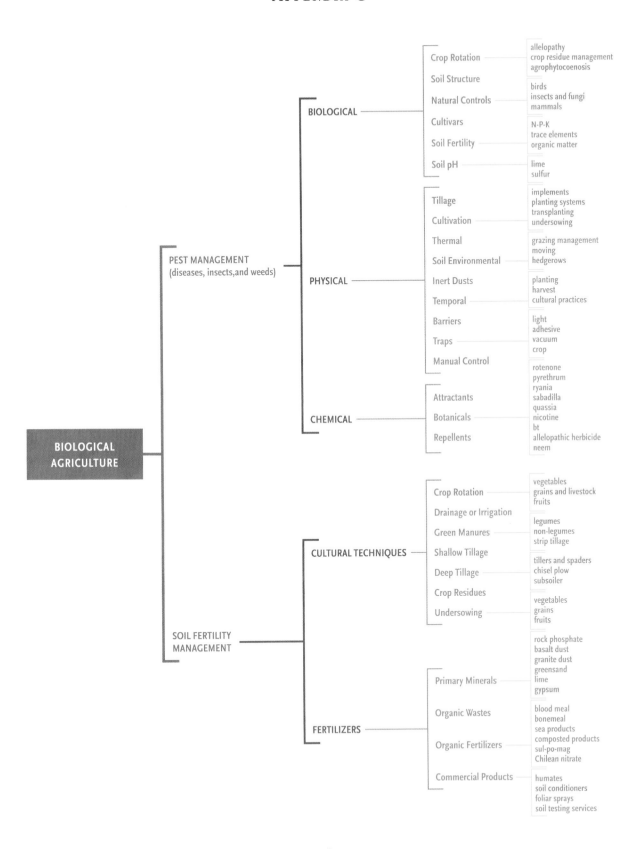

Recommended Tools and Suppliers

Tools have their own integrity;
The sneath of the scythe curves rightly to the hand,
The hammer knows its balance, knife its edge,
All tools inevitably planned,
Stout friends, with pledge
Of service; with their crochets too
That masters understand . . .

—V. SACKVILLE-WEST

When the first edition of this book was written in the 1980s, there was no Internet and no easy way to research and find the ideal tools and supplies for the small vegetable farm. Therefore, I added a detailed tools and supplies section with addresses and phone numbers. At the speed the world changes, lists like that quickly become out of date. So for this third edition, I heartily recommend the Internet and your favorite search engine. If you look diligently you should quickly find potential suppliers. Then carefully investigate them. I have, and I highly recommend Johnny's Selected Seeds as a dependable source for most all of the equipment that we use here on Four Season Farm.

Seeds

For many years I have been buying seeds with confidence from:

Johnny's Selected Seeds
955 Benton Avenue
Winslow, ME 04901
(877) 564-6697
www.johnnyseeds.com

High Mowing Organic Seeds
76 Quarry Road
Wolcott, VT 05680
(866) 735-4454
www.highmowingseeds.com

Fedco Seeds
PO Box 520
Clinton, ME 04927
(207) 426-9900
www.fedcoseeds.com

Harris Seeds
355 Paul Road
Rochester, NY 14624
(800) 544-7938
www.harrisseeds.com

Stokes Seeds
Box 548
Buffalo, NY 14240
(800) 396-9238
www.stokeseeds.com

Territorial Seed Co.
PO Box 158
Cottage Grove, OR 97424
(800) 626-0866
www.territorialseed.com

Caring for Tools

Keep them clean. Clean all tools after using them and before putting them away. Use a wooden or metal scraper to remove dirt, a wire brush for the finer material, and wipe dry with a rough rag.

Keep them sharp. A number of benefits accrue from keeping garden tools well sharpened. The work can be done better; the tools require less effort to use; more can be accomplished in a given time; and the worker will feel much less tired at the end of the job.

It is best to touch up the edge of a tool frequently to keep it sharp rather than waiting until it has become dull. For most horticultural tools, a flat metal file is the implement of choice for sharpening. A whetstone should be used for sickles and scythes. Knives and pruners are most effectively sharpened on an oilstone.

Keep them lively. Most good tools will have strong, straight-grained ash handles. These handles need to be coated occasionally with boiled linseed oil to prevent the wood from drying out. Dry wood loses the valuable resiliency of a properly maintained handle and is more likely to break.

Keep them around. One of the neatest, best-kept, and most efficient tool rooms I have ever seen was on a small farm in Germany. There was a prominent sign over the workbench that, translated into English, read: EVERYTHING IN ITS PLACE SAVES ANGER, TIME, AND WORDS.

Hang small tools on cup hooks or a pegboard. If the location is marked by a painted outline, the tool can be easily returned to its place after use and noticed when it is missing. Another worthwhile practice is to paint a conspicuous color (red, light blue, or orange) on a non-wearing area of the metal part of the tool, so it can be seen easily if it is left lying about the fields.

A well-planned toolshed should be constructed in close proximity to the fields. It should be equipped with electricity, running water, a concrete floor, a woodstove for heat, a workbench with vises, and adequate safe-storage facilities for gas, oil, parts, and tools.

ANNOTATED BIBLIOGRAPHY

Note: Wherever possible the publisher and date are given for the earliest edition of each book listed in this bibliography.

Contemporary Sources of Information

A common characteristic of many successful organic growers is that they learned what they know without outside help. They managed to decipher the whys and wherefores of biological systems on their own plucky initiative. This prevalence of self-education is not surprising, since so little specific instructional material or consultation was available years ago. In such a situation the best teachers, after experience, are good, basic, general reference sources on soils, plants, and techniques. The information can be interpreted to meet the grower's specific needs. These books are not light reading, but they do offer a wealth of useful ideas and can provide the "hard data" that serve as the springboard to improved performance.

Balls, R. *Horticultural Engineering Technology.* London and New York: Macmillan, 1985. This book, along with M. F. J. Hawker and J. F. Keenlyside's *Horticultural Machinery* and John Robertson's *Mechanising Vegetable Production* (see below), is valuable both as background on how things work and as a source of ideas for home-fabricated solutions to the same tasks on a smaller scale. As a handy old Maine neighbor used to tell me, "Wal, if you can give me an idea what you want, I 'spect I can gump something up for ya." Good inspirational books for "gumpers."

Bleasdale, J. K. A. *Plant Physiology in Relation to Horticulture.* London: Macmillan, 1984. It is always nice to know what is going on behind the scenes. This book has all the "inside" information. A lot heavier going than the Bleasdale books below, but well worth the trouble for the hard facts and excellent bibliographic references.

Bleasdale, J. K. A., P. J. Salter, et al. *Know and Grow Vegetables.* 2 vols. London and New York: Oxford University Press, 1979 and 1982. Wonderful little books for the amateur as well as the budding professional. The authors, from the National Vegetable Research Station in England, obviously like their subject and delight in providing the reader with first-class information, from both their experience and their experiments.

Bunt, A. C. *Modern Potting Composts.* University Park: Pennsylvania State University Press, 1976. To an Englishman like Bunt, *potting composts* is the term for "potting soils." This is my favorite of all the books I consulted on the subject. Everything you ever wanted to know.

Engeland, Ron L. *Growing Great Garlic.* Okanogan, WA: Filaree Productions, 1991. A highly enjoyable book about all aspects of garlic culture. Written in a down-home style by a grower for growers.

Facciola, Stephen. *Cornucopia II: A Source Book of Edible Plants.* Vista, CA: Kampong Publications, 1998. This is an almost unbelievable resource. It sources every seed of every edible plant from every seed company as of the date of publication. Also features an awesome bibliography. Very well done.

Flegmann, A. W., and Raymond A. T. George. *Soils and Other Growth Media.* London and New York: Macmillan, 1979. The heaviest of the bunch, but still worth having as a reference work.

Flint, Mary Louise. *Pests of the Garden and Small Farm.* Davis: University of California, 1990. Excellent color pictures and scale drawings for identification. Practical problem-solving information.

Fordham, R., and A. G. Biggs. *Principles of Vegetable Crop Production.* London: William Collins Sons (distributed in the United States by Sheridan House), 1985. Very complete and not overly dry. Lots of little tidbits of useful information tucked here and there.

Grower Guides: No. 3, *Peppers and Aubergines*; No. 7, *Plastic Mulches for Vegetable Production*; No. 10, *Blocks for Transplants*; No. 21, *Lettuce Under Glass*; No. 26, *Vegetables Under Glass.* Parts of a continuing series that used to be published by Grower Books, Nexus Media, Nexus House, Swanley, Kent, England BR8 8HY. These are consistently informative booklets, usually under 100 pages, that do a thorough job of treating each individual subject.

Hawker, M. F. J., and J. F. Keenlyside. *Horticultural Machinery.* New York: Longman, 1985. See my comments under R. Balls, page 261.

Lorenz, Oscar A., and Donald N. Maynard. *Knott's Handbook for Vegetable Growers*, 3rd ed. New York: Wiley Interscience, 1988. An indispensable reference. All growers should have a copy. The book contains almost every fact that, sooner or later, you may need to know.

Magdoff, Fred, and Harold van Es. *Building Soils for Better Crops.* 3rd ed. College Park, MD: Sustainable Agriculture Research and Education (SARE) program, 2009. www.sare.org/Learning-Center /Books/Building-Soils-for-Better-Crops-3rd -Edition. A valuable practical guide to ecological soil management.

Mastalertz, John W. *The Greenhouse Environment.* New York: John Wiley & Sons, 1977. A complete factual presentation of all aspects of greenhouse growing.

Mefferd, Andrew. *The Greenhouse and Hoophouse Growers Handbook.* White River Junction, VT: Chelsea Green, 2017. Excellent up-to-date information from a knowledgeable practitioner.

Ministry of Agriculture, Fisheries and Food. *Plant Physiological Disorders*, Reference Book 223. London: Her Majesty's Stationery Office, 1985. Problems do occasionally arise, even for the best of growers. When they do, books like this—along with J. B. D. Robinson's, and Roorda van Eysinga and Smilde's (below)—are nice "consultants" to have on hand. Clear color photos, detailed diagnoses, solid advice, and excellent references make these books a worthwhile investment.

Nelson, Paul V. *Greenhouse Operation and Management.* Reston, VA: Reston Publishing, 1978. A complete presentation of all aspects of greenhouse growing.

Parnes, Robert. *Fertile Soil.* Davis, CA: agAccess, 1990. A very informative book that is written for growers. Full of useful tables and appendices.

Robertson, John. *Mechanising Vegetable Production.* Ipswich, Suffolk, England: Farming Press, 1978. See my comments under R. Balls (above).

Robinson, D. W., and J. G. D. Lamb, eds. *Peat in Horticulture.* New York: Academic Press, 1975. Everything you ever wanted to know about peat, along with all its uses and horticultural qualities.

Robinson, J. B. D., ed. *Diagnosis of Mineral Disorders in Plants.* Vol. 1, *Principles*; Vol. 2, *Vegetables*; Vol. 3, *Glasshouse Crops.* London: Her Majesty's Stationery Office, 1983, 1983, and 1987.

Roorda van Eysinga, J. P. N. L., and K. W. Smilde. *Nutritional Disorders in Glasshouse Tomatoes, Cucumbers and Lettuce.* Wageningen, The Netherlands: Centre for Agricultural Publishing and Documentation, 1981.

Sarrantonio, Marianne. *Northeast Cover Crop Handbook.* Emmaus, PA: Rodale Institute, 1994. An excellent source of information and suggestions on how to assess the effects of cover crops.

Taylor's Guide to Vegetables & Herbs. Boston: Houghton Mifflin, 1987. The color photos are superb. If you don't know what a particular vegetable looks like (sorrel, arugula, orach, Malabar spinach, et cetera), this book will show you.

Tite, R. L. *Growing Tomatoes: A Greenhouse Guide*. London: Her Majesty's Stationery Office, 1983. This is a British ADAS (extension) publication. A very well-done, small (32 pages) introductory booklet that will help any greenhouse tomato grower get off on the right foot.

Wittwer, S. H., and S. Honma. *Greenhouse Tomatoes, Lettuce and Cucumbers*. East Lansing: Michigan State University Press, 1979. A shade dated but still good basic information. This is the book I started with.

Classic Sources for the Organic Grower

Here is a listing of those "classic" sources that I believe most merit the attention of serious growers. Many of these are out of print, especially the classic English sources. However, some of these titles have been recently reprinted. Most of the rest can be found in a good library or through interlibrary loan.

As you might expect, the authors of these books do not always agree with one another. Some have written on the periphery of biological agriculture, while others were deeply involved and knowledgeable practitioners. It is important to read critically, check references, compare, and see what the other side has to say in order not to become, like so many proselytizers of a new idea, "a man of vast and varied misinformation."

Feeding quality is the most important matter of all. If a plant is healthy, and growing up to its own perfection, it must have great vitality, and it is the vitality, the living force of the plant, that heightens its food value. A vegetable can not give what it has not got; what it has, it gets from the soil. It cannot reach its "own perfection" in starved ground, still less in ground doped with chemicals.

—**Maye Bruce**, *Common-Sense Compost Making*, 1967

It would save much confusion if we all adopted the name "biological farming" rather than "organic farming." We should then keep the emphasis where it belongs, on the fostering of life and on biological balance, and not on just one of the techniques for achieving this, which, if narrowly interpreted, may be effective only in a certain set of circumstances.

—**Lady Eve Balfour**, *Journal of the Soil Association*, January 1954

Albrecht, William A. *Soil Fertility and Animal Health*. Webster City, IA: Fred Hahne, 1958. An outstanding survey of the subject by the most respected American exponent of intelligent farming. Albrecht begins with an old quote, "All flesh is grass," and proceeds to demonstrate the importance of the quality of that grass to animal health.

Aubert, Claude. *L'Agriculture Biologique*. Paris: LeCourrier du Livre, 1970. An able presentation of the case by a leading European expert. In French.

Baker, C. Alma. *Peace with the Soil*. Federated Malay States: self-published, 1939. An early exposition of the importance of feeding the soil rather than feeding the plant.

———. *The Labouring Earth*. London: Heath Cranton, 1940. A survey of agriculture from the biodynamic point of view. A case could be made for Baker having used the word *organic* as early as Northbourne.

Balfour, E. B. (Lady Eve). *The Living Soil*. London: Faber, 1943. The important early work by a founder of the Soil Association. Lady Eve documents the evidence for biological agriculture. A fine book that should be in everyone's library. Reprinted 1975.

Billington, F. H. *Compost for Garden Plot or Thousand Acre Farm*. London: Faber, 1943. An early work giving thorough treatment to all aspects of composting. Five specific methods are described in detail. There is also a more recent edition revised and co-authored by Ben Easey.

> As always in my experience, the destructive activity of insects came only when plants were in an abnormally weak condition.
>
> —**Edward H. Faulkner**, *Soil Restoration*, 1953

Blake, Michael. *Concentrated Incomplete Fertilizers.* London: Crosby Lockwood, 1966. A discussion of the faults and consequent abuses of chemical fertilizers.

Bromfield, Louis. *Pleasant Valley.* New York: Harper, 1946. This is the first of Bromfield's farming books. In it, he relates how he returned to Ohio and became a farmer, and discusses the details of early farm plans, soil conservation, and the Friends of the Land.

———. *Malabar Farm.* New York: Harper, 1947. Continues the story begun by *Pleasant Valley* and covers the year-round rhythm of activities at Malabar. Also focuses on Bromfield's other interests with such chapters as "Grass the Great Healer," "Malthus Was Right," and "The Organic-Chemical Fertilizer Feud."

———. *Out of the Earth.* New York: Harper, 1948. Stresses the need for knowledge of the many intricate, interrelated sciences involved in agriculture as a complement to the knowledge of the farm itself. Bromfield condemns the idea that "anybody can farm." Practical intelligence and dedication are necessary for success.

———. *From My Experience.* New York: Harper, 1955. The last of the farm books and the best of the lot. Outstanding accounts of a roadside market, farming in Brazil, building topsoils, living with the weather, and a chapter titled "A Hymn to Hawgs" make enjoyable and informative reading.

Bruce, Maye. *From Vegetable Waste to Fertile Soil.* London: Faber, 1940.

———. *Common-Sense Compost Making.* London: Faber, 1967. Both of Bruce's books describe composting with the aid of herbal extracts. The extracts supposedly activate the heap and produce a superior finished product. The standard work. Recently reprinted.

Burr, Fearing. *The Field and Garden Vegetables of America.* Chillicothe, IL: The American Botanist, Booksellers, 1994. A classic work, first published in 1863. Contains over 600 pages describing nearly 1,100 garden varieties grown more than 150 years ago; illustrated with woodcuts.

Cocannouer, Joseph A. *Weeds: Guardians of the Soil.* New York: Devin-Adair, 1952. Cocannouer is an enthusiastic advocate of the virtues of weeds.

———. *Farming with Nature.* Norman: University of Oklahoma Press, 1954. A general work with some good information. Reprinted under the title *Organic Gardening and Farming.*

———. *Water and the Cycle of Life.* New York: Devin-Adair, 1958. A searing indictment of the mistaken farming practices that led to the Dust Bowl and their effect on the ecology of water.

Corley, Hugh. *Organic Farming.* London: Faber, 1957. "But the reason for farming well is that it is right." Corley fills this book with useful interpretations of what "farming well" is all about.

Darwin, Charles. *The Formation of Vegetable Mould Through the Actions of Worms.* London: Faber and Faber, 1945. This timeless classic has been republished with an enthusiastic introduction by Sir Albert Howard.

Donaldson, Frances. *Approach to Farming.* London: Faber, 1941. This book states that the "health" of

> It is the same with almost everything; we studied, compared, and observed before attempting it. Somewhere there is always someone who is doing a job a little better and there are many who are doing it a great deal worse; from either a lot can be learned.
>
> —**George Henderson**, *The Farming Ladder*, 1944

the soil, of the livestock, and of the produce is the paramount consideration on any farm.

Easey, Ben. *Practical Organic Gardening*. London: Faber, 1955. An outstanding work, almost a textbook. Very thorough and well documented. Contains a lot of material found nowhere else.

Elliot, Robert H. *The Clifton Park System of Farming*. London: Faber, 1907. In his introduction, Sir George Stapledon calls this book an "agricultural classic." First published in 1898 under the title *Agricultural Changes*, it was later the work that inspired Sykes and Turner. Elliot writes of grass, pasture, and especially of his extensive seed mixture, "calculated to fill the land with vegetable matter."

Faulkner, Edward H. *Plowman's Folly*. Norman: University of Oklahoma Press, 1943. Louis Bromfield wrote that everyone including Hollywood actresses asked him about this book. It ultimately sold millions of copies. An effective condemnation of the moldboard plow.

———. *A Second Look*. Norman: University of Oklahoma Press, 1947. In this book Faulkner attempts to restate his case more clearly in view of the controversy stirred up by *Plowman's Folly*.

———. *Soil Restoration*. London: Michael Joseph, 1953. Faulkner applied his techniques to bring a worn-out farm back into production as a market garden. This is the story of that experiment.

Godwin, George. *The Land Our Larder*. London: Acorn Press, 1940. The story of one of the first English farms following the ideas of Sir Albert Howard and its transformation into one of the most fertile and flourishing agricultural enterprises in the country.

Graham, Michael. *Soil and Sense*. London: Faber, 1941. An unpretentious but informative little book about grasses, pastures, livestock, and their relationships to one another.

Hainsworth, P. H. *Agriculture: A New Approach*. London: Faber, 1954. A very reasonable and well-documented study of biological agriculture

Good soil management, in the sense of maintaining fertility and productivity, depends upon a number of relatively simple practices: (1) Suitable tillage; (2) maintaining the supply of organic matter, principally by the use of rotations and cover crops including legumes; (3) correcting soil acidity in humid regions; (4) providing an adequate supply of phosphorus; and (5) using mechanical methods to control erosion where rotation and cover cropping are not sufficient.

— *Soils and Men*, USDA Yearbook
of Agriculture 1938

by a successful market grower. Contains a lot of new and stimulating material. One of my personal favorites.

Hambidge, Gove, ed. *Soils and Men: Yearbook of Agriculture 1938*. USDA. Washington, DC: US Government Printing Office, 1938. A most amazing book that could be reprinted today as an organic farming textbook. The science of agriculture in the 1930s (before the chemical invasion) was very much in line with the principles of organic farming.

Henderson, George. *The Farming Ladder*. London: Faber, 1944.

———. *Farmer's Progress*. London: Faber, 1950.

———. *The Farming Manual*. London: Faber, 1960. If you only read one inspirational author on farming, read Henderson. The first two books cover his entry into farming with his brother and their experience over the years. The third is a detailed guide to farmwork. Henderson infuses all these books with his own love of farming and an invaluable sense of craftsmanship and pride in a job done well.

Henderson, Peter. *Gardening for Profit*. New York: Orange Judd, 1867. Peter Henderson was the market-gardening authority of his time, and he sure knew his stuff. Even after 150 years his

By 1919 I had learnt how to grow healthy crops, practically free from diseases, without the slightest help from mycologists, entomologists, bacteriologists, agricultural chemists, statisticians, clearing-houses of information, artificial manures, spraying machines, insecticides, fungicides, germicides, and all the other expensive paraphernalia of the modern experiment station. This preliminary exploration of the ground suggested that the birthright of every crop is health.

—**Sir Albert Howard**, *The Soil and Health*, 1947

advice still rings true in most particulars. Recently reprinted in a new edition by the American Botanist Booksellers of Chillicothe, IL.

Hills, Lawrence. *Russian Comfrey*. London: Faber, 1953. Comfrey is a perennial crop used for feed, mulching, and compost. This book details many useful ways of employing comfrey in the farm economy.

———. *Down to Earth Fruit and Vegetable Growing*. London: Faber, 1960. With typical thoroughness Lawrence Hills, founder of the Henry Doubleday Research Association, covers every aspect of the garden with straightforward, practical, and detailed instructions.

Hopkins, Cyril G. *Soil Fertility and Permanent Agriculture*. Boston: Ginn and Company, 1910. This is Hopkins's best-known work and his most thorough exposition of the concept of a "permanent agriculture."

Howard, Sir Albert. *An Agricultural Testament*. London: Oxford University Press, 1940. The most important, indeed the seminal work of biological agriculture, it inspired countless readers to try Howard's ideas. The book presents ways and means by which the fertility of the soil can be restored, maintained, and improved by natural methods.

———. *The Soil and Health*. New York: Devin-Adair, 1947. A continuation of the ideas of *An Agricultural* *Testament*, presented in a more popular form. "I have not hesitated to question the soundness of present-day agricultural teaching and research . . . due to failure to realize that the problems of the farm and garden are biological rather than chemical."

Howard, Albert, and Yeshwant D. Wad. *The Waste Products of Agriculture*. London: Oxford University Press, 1931. Howard's first book about his compost-making experiences in India.

Howard, Louise. *The Earth's Green Carpet*. London: Faber, 1947. A popular recounting of the ideas of Sir Albert Howard through the eyes of his wife. Well done.

———. *Sir Albert Howard in India*. London: Faber, 1953. Traces the development of Howard's thought during his years as a researcher in India. A valuable record of his scientific work.

Hunter, Beatrice T. *Gardening Without Poisons*. Boston: Houghton Mifflin, 1964. Undoubtedly the best-documented and most thoroughly researched work on the subject. Well organized, with an excellent index and bibliography.

Jacks, G. V., W. D. Brind, and Robert Smith. *Mulching*. Technical Communication No. 49 of the Commonwealth Bureau of Soil Science, 1955. A very complete study of different types of mulching and their effects of soil fertility and soil creatures.

Jacks, G. V., and R. O. Whyte. *Alternate Husbandry*. Imperial Agricultural Bureau, Joint Publication No. 6, Great Britain, May 1944. A scientific presentation of all aspects of ley farming.

Jenks, Jorian. *The Stuff Man's Made Of*. London: Faber, 1959. The origin, the philosophy, and the scientific evidence behind biological agriculture. Jenks, for many years editor of the *Journal of the Soil Association*, has an encyclopedic grasp of the subject.

King, F. C. *The Compost Gardener*. Highgate, Kendal, Lancashire, England: Titus Wilson, 1943. This small book lays down the general principles of cultivation for all the popular vegetables. Contains some unique information.

———. *Gardening with Compost*. London: Faber, 1944. Compost preparation and use, comments on chemical fertilizers, and sections on weeds and earthworms.

———. *The Weed Problem*. London: Faber, 1951. King is doubly unorthodox. He defends the control rather than elimination of weeds, and he condemns turning over the soil.

King, F. H. *Farmers of Forty Centuries*. London: Jonathan Cape, 1927. One of the granddaddies of them all, this classic was first published in 1911. King's trip through China, Korea, and Japan showed him how soil fertility had been preserved by returning all organic wastes to the land. Hundreds of photos and fascinating information. Recently reprinted.

Konnonova, M. M. *Soil Organic Matter*. New York: Pergamon, 1961. A technical work well worth reading for a better understanding of the processes involved in biological agriculture.

Krasil'nikov, N. A. *Soil Microorganisms and Higher Plants*. Washington: U.S. Department of Commerce, 1958. An extensive compendium of useful information about life in the soil. Excellent chapter titled "Biological Factors of Soil Fertility." "The active principles of humus and composts are not the mineral nutrients present in them, but the organic substances and the biologically active metabolites of microbes."

Lawrence, W. J. C. *Catch the Tide*. London: Grower Books, 1980. The horticultural adventures and investigations of W. J. C. Lawrence. Great reading.

Lawrence, W. J. C., and J. Newell. *Seed and Potting Composts*. London: Allen & Unwin, 1939. Explains much of the early work done to develop better potting mixes.

Maunsell, J. E. B. *Natural Gardening*. London: Faber, 1958. A book of unconventional gardening techniques. Maunsell is the most thorough of the no-diggers, and his use of the spading fork for "disturbing" the soil is worth noting.

Morris, Edmund. *Ten Acres Enough*. New York: American News, 1864. Subtitled "How a Very Small Farm May Be Made to Keep a Very Large Family." Morris's advice is as sound today as it was then.

Northbourne, Lord. *Look to the Land*. London: Dent, 1940. One of the early inspirational works. "Mixed farming is economical farming, for only by its practice can the earth be made to yield a genuine increase." Northbourne was one of the first, if not the first, to call it "organic" farming.

O'Brien, R. Dalziel. *Intensive Gardening*. London: Faber, 1956. This book of original ideas describes a meticulously efficient market garden. From the layout, to the philosophy, to the composting and fertilizing procedures—even to a motion study of transplanting—everything is covered. Veganic compost (without animal manure) is used. Another favorite of mine.

Oyler, Philip. *The Generous Earth*. London: Hodder and Stoughton, 1950. Tells the story of the timeless farm life in the Dordogne Valley of France, "the land of all good things." It shows how the operation of sound farming practices will sustain fertility indefinitely.

———. *Sons of the Generous Earth*. London: Hodder and Stoughton, 1963. More on Oyler's experience

> We are not out to convince anyone of the truth of the discoveries we have made of the way the soil transforms itself in three years using our methods, for we are confident ourselves that time will do that for us. We put forward this method as an alternative to the orthodox gardening techniques, which to-day involve growers in heavy labour costs and outlay on stable and artificial manures, things which bite so deeply into the profits of intensive cultivation of vegetables and plants. When we describe how something should be done, we have done it that way and made a profit out of it.
>
> —**Dalziel O'Brien**, *Intensive Gardening*, 1956

in France. An important story from a man who values hard work, rural skills, wholesome food and drink, and a simpler way of life.

Pfeiffer, Ehrenfried. *Bio-Dynamic Farming and Gardening*. New York: Anthroposophic Press, 1938. Presents the case for non-chemical farming in general, and biodynamic farming in particular.

———. *The Earth's Face and Human Destiny*. Emmaus, PA: Rodale, 1947. A discussion of landscape characteristics and their value to the natural system.

Picton, Dr. Lionel. *Nutrition and the Soil*. New York: Devin-Adair, 1949. Mostly on nutrition, but partly about the soil. One of the earliest works on the subject and therefore of some historical interest.

Poore, George Vivian. *Essays on Rural Hygiene*. London: Longmans, Green, 1903. A highly influential early work with a chapter titled "The Living Earth."

Rayner, M. C. *Problems in Tree Nutrition*. London: Faber, 1944. A report of the work done by Dr. Rayner at Wareham Heath. The use of composts in forestry to encourage the growth of seedlings in a sterile soil by stimulating the development of mycorrhizal associations.

Rodale, J. I. *Stone Mulching in the Garden*. Emmaus, PA: Rodale, 1949. An almost forgotten work and one of Rodale's best. Mulching with stones, an old and effective practice, is clearly explained in photos and text.

———. *Encyclopedia of Organic Gardening*. Emmaus, PA: Rodale, 1959.

———. *How to Grow Vegetables and Fruits by the Organic Method*. Emmaus, PA: Rodale, 1960. This and the encyclopedia above are large (1,000-page) books covering all phases of the art. This is the first book I read.

Rowe-Dutton, Patricia. *The Mulching of Vegetables*. Farnham Royal, Buckinghamshire, England: Commonwealth Agricultural Bureaux, 1957. A valuable compilation of all the research on mulching up to the date of publication.

Royal Horticultural Society. *The Vegetable Garden Displayed*. London, 1942. This World War II classic does a great job teaching vegetable growing with 300 black-and-white photographs.

Russell, E. J. *The World of the Soil*. London: Collins, 1957. A thorough study of the soil by a director of the Rothamsted Experimental Station in England. Reliable background information for anyone.

Seifert, Alwin. *Compost*. London: Faber, 1952. An outstanding book on the hows and whys of producing and using first-class compost.

Shewell-Cooper, W. E. *The Complete Fruit Grower*. London: Faber, 1960.

———. *The Complete Vegetable Grower*. London: Faber, 1968. Encyclopedic coverage of both subjects in a readable format.

Smith, Gerard. *Organic Surface Cultivation*. London: Faber, 1961. Another of the no-digging books. Deals with composts, garden planning, plus an assortment of hints and ideas.

Soil Association, The. *Journal of the Soil Association*, 1947–1972. Walnut Tree Manor, Haughley, Stowmarket, Suffolk IP14 3RS, England. A quarterly journal of invaluable reference information.

Stapledon, R. George, and William Davies. *Ley Farming*. London: Faber, 1941. The best book on the subject. "Considered as an agent for the promotion of soil fertility, the grass sod, properly managed and intelligently converted, must be regarded as perhaps the most valuable foundation upon which the farmer can build."

I remember the time when the stable would yield,
Whatsoever was needed to fatten a field.
But chemistry now into tillage we lugs
And we drenches the earth with a parcel of drugs.
All we poisons, I hope, is the slugs.

— ***Punch***, 1846, as quoted in the
Journal of the Soil Association, April 1956

Stephenson, W. A. *Seaweed in Agriculture and Horticulture*. London: Faber, 1968. Documents the use of seaweed—especially in liquefied form—in farming, by examples of research from various parts of the world. Those interested in the subject will find some additional information in *Seaweed Utilization* by Lily Newton (London: Sampson-Low, 1951) and *Seaweeds and Their Uses* by V. J. Chapman (London: Methuen, 1970).

Sykes, Friend. *Humus and the Farmer*. London: Faber, 1946. The transformation of unpromising land into one of the showplace farms of England by methods described as humus farming. Covers renovating old pastures, making new ones, subsoiling, harvesting, and related topics.

———. *Food, Farming and the Future*. London: Faber, 1951. The further development of humus farming plus many peripheral subjects.

———. *Modern Humus Farming*. London: Faber, 1959. Discusses the danger to the soil caused by worship of "technical efficiency" and "getting more for less." Sykes puts forth his case that humus farming is as effective and productive as any other system.

Turner, Newman. *Fertility Farming*. London: Faber, 1951. Turner was a practical farmer who learned conventional agriculture in college, but when he applied the teachings the results were disastrous. He then "unlearned" all his formal training and formulated his own system. Fascinating reading.

———. *Herdsmanship*. London: Faber, 1952. Dedicated "to the Jersey Cow, which combines beauty with efficiency." Comprehensive treatment of dairy cow selection and management from Turner's point of view. Excellent descriptions of all major dairy breeds plus sections on herbal veterinary practices.

———. *Fertility Pastures*. London: Faber, 1955. The value of the herbal ley (temporary pasture) is the central theme of this book. The detailed information on the character and properties of herbs and grasses for grazing is extremely interesting.

The addition of large amounts of organic matter, especially fresh plant and animal residues, to the soil completely modifies the nature of its microbiological population. The same is true of changes in soil reaction which are brought about by liming or by the use of acid fertilizers, by the growth of specific crops, notably legumes, and by aeration of soil resulting from cultivation.

—**Selman Waksman**, *Soil Microbiology*, 1952

Turner determined the composition of his pasture seed mixtures by "consulting the cow."

Voisin, Andre. *Grass Productivity*. London: Crosby Lockwood, 1958.

———. *Soil, Grass and Cancer*. London: Crosby Lockwood, 1959.

———. *Better Grassland Sward*. London: Crosby Lockwood, 1960. Voisin, a leading French authority on grassland management, was deeply concerned with the biological quality of produce. These works inspired the modern interest in rotational grazing.

Waksman, Selman A. *Soil Microbiology*. New York: John Wiley, 1952. An extremely valuable book. Waksman details the needs of soil microorganisms and their importance in the soil. His information is consistent with the best practices of biological agriculture.

Weaver, John E. *Root Development of Field Crops*. New York and London: McGraw-Hill, 1926.

Weaver, John E., and William E. Bruner. *Root Development of Vegetable Crops*. New York and London: McGraw-Hill, 1927. Weaver and his crew investigated crop root systems by digging deeply to extract them from the earth. Absolutely fascinating information about what is happening below the surface of the soil.

Whyte, R. O. *Crop Production and Environment*. London: Faber, 1960. A book of plant ecology. It treats the effects on the plant of what Whyte considers to be the primary factors of aerial environment: temperature, light, and darkness.

Wickenden, Leonard. *Make Friends with Your Land*. New York: Devin-Adair, 1949. Wickenden was a professional chemist who became fascinated by organic growing. In this book he attempts to cut through some of the myths and to investigate the claims from a scientific perspective. An exceptional book for the skeptical beginner.

————. *Gardening with Nature*. New York: Devin-Adair, 1954. A thoroughly professional book for beginner and experienced gardener alike. A

practical and comprehensive treatment of all aspects of gardening.

————. *The Wheel of Health*. London: C. W. Daniel, 1938. What is health and what is the relation of food quality to soil quality.

Wrench, G. T. *Reconstruction by Way of the Soil*. London: Faber, 1946. A historical survey of soil mistreatment and its influence on civilization from earliest times. Wrench views farming as a creative art.

Wright, D. Macer. *Fruit Trees and the Soil*. London: Faber, 1960. Soil management in the orchard as the key to better-quality fruit.

NOTES

Chapter 2: A Little History to Begin

1. Gunter Vogt, "The Origins of Organic Farming," in *Organic Farming: An International History*, edited by William Lockeretz (Wallingford, Oxfordshire, England: CAB International, 2007).

2. G. Vivian Poore, *Essays on Rural Hygiene* (London: Longmans, Green, 1893), 163.

3. Cyril G. Hopkins, *Soil Fertility and Permanent Agriculture* (Boston: Ginn and Company, 1910).

4. Cyril G. Hopkins, *Shall We Use "Complete" Commercial Fertilizers in the Corn Belt?* (University of Illinois AES Circular No. 165, 1912), 1–20.

5. Vogt, "The Origins of Organic Farming," 15–16.

6. Vogt, "The Origins of Organic Farming," 18.

7. Pierre Delbet, *L'agriculture et la sante* (Paris, Denoél, 1945).

8. A. de Saint Henis, *Guide pratique de culture biologique* (Angers, France: Agriculture et Vie, 1972).

9. E. B. Balfour, *The Living Soil* (London: Faber and Faber, 1943).

10. Sir Albert Howard, *An Agricultural Testament* (London: Oxford University Press, 1940).

11. Karl B. Mickey, *Health from the Ground Up* (Chicago: International Harvester, 1946).

12. E. C. Auchter, "The Interrelation of Soils and Plant, Animal and Human Nutrition," *Science* 89, no. 2315 (1939): 421–27.

13. R. C. Barron, ed., *The Garden and Farm Books of Thomas Jefferson* (Golden, CO: Fulcrum, 1987), 169.

14. E. Darwin, *Philosophy of Agriculture and Gardening* (London: J. Johnson, 1800).

15. Thomas Green Fessenden, *The New American Gardener*, 13th ed. (Boston: Otis, Broaders, 1839), 169.

16. Vincent Gressent, *Le potager moderne*. (Paris: Librairie Agricole de la Maison Rustique, 1926), 135, 861–62.

17. Paul Sorauer, *Handbuch der pflanzenkrankheiten* (Berlin: P. Parey, 1905).

18. H. Marshall Ward, "On Some Relations Between Host and Parasite in Certain Diseases of Plants" (Croonian Lecture), *Proceedings of the Royal Society* 47 (1890): 303–433.

19. Kenneth F. Baker and R. James Cook, *Biological Control of Plant Pathogens* (San Francisco: W. H. Freeman, 1974).

20. Von H. Thiem, "Über Bedingungen der Massenvermehrung von Insekten," *Arbeiten über physiologische und angewandte entomologie aus Berlin Dahlem* 5, no. 3 (1938): 229–55.

21. Louise E. Howard, *Sir Albert Howard in India* (London: Faber and Faber, 1953), 162.

22. Sir Albert Howard, *An Agricultural Testament*, 161.

23. Fred Magdoff and Ray R. Weil, *Soil Organic Matter in Sustainable Agriculture* (Boca Raton, FL: CRC Press, 2004), 3.

24. Luciano Pasqualoto Canellas et al., "Humic Acids Isolated from Earthworm Compost Enhance Root Elongation, Lateral Root Emergence, and Plasma Membrane H+-ATPase Activity in Maize Roots," *Plant Physiology* 130 (2002): 1951–57.

25. Harry A. J. Hoitink and P. C. Fahy, "Basis for the Control of Soil Borne Plant Pathogens with Compost," *Annual Review of Phytopathology* 24 (1986): 93–114.

26. T. C. R. White, "The Abundance of Invertebrate Herbivores in Relation to the Availability of Nitrogen in Stressed Food Plants," *Oecologia* 63 (1984): 90–105.

27. Eliot W. Coleman and Richard L. Ridgeway, "Role of Stress Tolerance in Integrated Pest Management," in *Sustainable Food Systems*, edited by Dietrich Knorr (Westport, CT: AVI Publishing, 1983), 127.

28. Vinod Kumar et al., "An Alternative Agriculture System Is Defined by a Distinct Expression Profile of Select Gene Transcripts and Proteins," *Proceedings of the National Academy of Sciences of the United States of America* 101, no. 29 (2004): 10535–40.

29. Brian Halweil, *Still No Free Lunch* (Boulder, CO: Organic Center, 2004).

30. Benjamin D. Walsh, *The Practical Entomologist* (Philadelphia: Entomological Society of Philadelphia, 1866).

31. Thomas. B. Colwell Jr., "Some Implications of the Ecological Revolution for the Construction of Value," in *Human Values and Natural Science*, edited by E. Laszlo and J. B. Wilbur (New York: Gordon and Breach, 1970), 247.

Chapter 3: Agricultural Craftsmanship

1. George Henderson, *The Farming Ladder* (London: Faber & Faber, 1944).

2. Henderson, *The Farming Ladder*.

Chapter 4: Land

1. See Norman Rosenberg, Blaine Blad, and Shashi Verma, *Microclimate: The Biological Environment* (New York: John Wiley & Sons, 1983).

2. Two useful books for background on irrigation methods are Melvyn Kay's *Sprinkler Irrigation: Equipment and Practice* (London: Batsford, 1984) and Robert Kourik's *Drip Irrigation for Every Landscape and All Climates* (Santa Rosa, CA: Metamorphic Press, 1992).

3. I learned about this from Jeff Beringer, Fish & Wildlife Research Center, Missouri Department of Conservation, 1110 South College Avenue, Columbia, MO 65201; telephone (314) 882-9880.

Chapter 7: Marketing Strategy

1. Some books that deal with the ideas of food quality as it relates to the use of fertilizers and pesticides are: *Silent Spring* by Rachel Carson (Boston: Houghton Mifflin, 1962); *Nutritional Values in Crops and Plants* by Werner Schuphan (London: Faber, 1961); *Mineral Nutrition and the Balance of Life* by Frank Gilbert (Norman: University of Oklahoma Press, 1957); *Nutrition and the Soil* by Dr. Lionel Picton (New York: Devin-Adair, 1949); and *The Living Soil* by E. B. Balfour (London: Faber, 1943).

Chapter 8: Planning and Observation

1. A good book for those getting started in greenhouse vegetable production is *Greenhouse Tomatoes, Lettuce and Cucumbers* by S. H. Wittwer and S. Honma (East Lansing: Michigan State University Press, 1979).

2. This is the same layout a tractor-scale grower would refer to as a bed system.

3. Further information on succession planting of greenhouse lettuce can be found in *Lettuce Under Glass*, Grower Guide No. 21, published by Grower Books, 50 Doughty Street, London WCIN 2LP.

Chapter 9: Crop Rotation

1. The most complete bibliography of studies on all aspects of crop rotations up through 1975 is collected in G. Toderi, "Bibliografia sull avvicendamento delle colture," *Rivista di agronomia* 9 (1975): 434–68.

2. The results of the University of Rhode Island crop-rotation experiments are reported in the following two studies: "A Half-Century of Crop-Rotation Experiments" by R. S. Bell, T. E. Odland, and A. L. Owens in Bulletin No. 303 (1949), and "The Influence of Crop Plants on Those Which Follow: V" by T. E. Odland, R. S. Bell, and J. B. Smith in Bulletin No. 309 (1950) (Kingston: Rhode Island Agricultural Experiment Station).

Chapter 10: Green Manures

1. The most comprehensive book on green-manuring is still the classic *Green Manuring* by Adrian J. Pieters (New York: John Wiley, 1927). Obviously, good basic information is never out of date.

2. For a complete discussion of the decomposition of green manures in the soil, see *Soil Microbiology* by Selman A. Waksman (New York: John Wiley, 1952). *Soil Microorganisms and Higher Plants* by N. A. Krasil'nikov (1958), which is available from the US Department of Commerce, 5285 Port Royal Road, Springfield, VA 22161, provides an amazing wealth of information about the effects of different crops on soil properties.

3. Waksman, *Soil Microbiology*, 256.

4. A. R. Weinhold et al., "Influence of Green Manures and Crop Rotation on Common Scab of Potato," *American Potato Journal* 41, no. 9 (1964): 265–73.

Chapter 11: Tillage

1. An excellent bibliography is the "Annotated Bibliography on Soil Compaction" available from the American

Society of Agricultural Engineers, 2950 Niles Road, Saint Joseph, MI 49085-9659.

2. J. N. Collins, "Hoeman's Folly," *The Land* 4, no. 4 (1945): 445.

3. Aref Abdul-Baki and John Teasdale, *Sustainable Production of Fresh-Market Tomatoes with Organic Mulches,* USDA Agricultural Research Service Farmers Bulletin FB-2279 (1994).

4. Anonymous, "Good Crops and an End to Soil Damage," *Ecos* 69 (Spring 1991): 1–6.

Chapter 12: Soil Fertility

1. Powdered or pelleted soil amendments need to be spread evenly. At one time or another, I have used the following methods: (1) Mix them together with the compost or manure and spread, or (2) spread with a metered spreader, such as the Gandy, which can be adjusted for accurate coverage. I recommend the latter. A 4-foot- (1.2 meter) wide Gandy spreader is easy to pull or tow with the walking tractor and allows for accurate spreading at almost any rate.

    For very light applications of trace elements, use a hand-cranked, chest-mounted seeder filled with the soil amendment rather than seed.

2. Your state geologist or department of mines can usually provide you with a list of quarries or rock-crushing operations in your area that may have finely ground rock powder wastes (called trap float at my local basalt quarry). These are often available free for the hauling. Ask for an analysis to make sure there are no undesirable contaminants.

3. A large number of studies have been done. Two are representative: A. S. Cushman, "The Use of Feldspathic Rocks as Fertilizers," USDA Bureau of Plant Industry Bulletin No. 104 (1907); and W. D. Keller, "Native Rocks and Minerals as Fertilizers," *Scientific Monthly* 66 (February 1948): 122–30.

4. C. C. Lewis and W. S. Eisenmenger, "Relationship of Plant Development to the Capacity to Utilize Potassium in Orthoclase Feldspar," *Soil Science* 65 (1948): 495–500.

5. Cyril G. Hopkins, *Shall We Use "Complete" Commercial Fertilizers in the Corn Belt?* (University of Illinois AES Circular No. 165, 1912), 1–20.

Chapter 13: Farm-Generated Fertility

1. D. Fritz and F. Venter, "Heavy Metals in Some Vegetable Crops as Influenced by Municipal Waste Composts," *Acta Horticulturae* 222 (1988): 51–62.

2. Arthur B. Beaumont, *Artificial Manures* (New York: Orange Judd, 1943); George Bommer, *New Method Which Teaches How to Make Vegetable Manure* (New York: Redfield and Savage, 1845).

3. The following study reinforces my experience: Andrew G. Hashimoto, "Final Report: On Farm, Composting of Grass Straw," Oregon State University, 1993.

4. An excellent instructional manual is *The Scythe Book* by David Tresemer (Brattleboro, VT: By Hand & Foot, 1981).

5. Some background on this topic can be found in P. M. Huang and M. Schnitzer, eds., "Interactions of Soil Minerals with Natural Organics and Microbes," SSSA Special Publication No. 17 (Madison, WI: Soil Science Society of America, 1986).

6. M. M. Mortland, A. E. Erickson, and J. E. Davis, "Clay Amendments on Sand and Organic Soils," *Michigan State University Quarterly Bulletin* 40, no. 1 (1957): 23–30. A more recent study confirms those results: Gerhard Reuter, "Improvement of Sandy Soils by Clay-Substrate Application," *Applied Clay Science* 9 (1994): 107–20.

Chapter 14: The Self-Fed Farm

1. *Alternate Husbandry,* May 1944, Imperial Agricultural Bureau, Joint Publication No. 6.

Chapter 16: Transplanting

1. The best source for information on all topics such as germination temperatures, numbers of seeds or transplants required for a given acreage, and any agricultural tables and lists ever compiled is *Knott's Handbook for Vegetable Growers,* 3rd ed., by Oscar A. Lorenz and Donald N. Maynard (New York: John Wiley & Sons, 1988).

Chapter 17: Soil Blocks

1. Michael D. Coe, "The Chinampas of Mexico," *Scientific American* (July 1964).

2. W. J. C. Lawrence, *Catch the Tide: Adventures in Horticultural Research* (London: Grower Books, 1980), 73–74.

3. The old-time official definition of *loam* referred to the crumbly, dark brown "soil" made by stacking layers of

sod from a fertile grass field upside down to decompose for a year or two. The development of the old loam-based mixes is well covered in *Seed and Potting Composts* by W. J. C. Lawrence and J. Newell (London: Allen & Unwin, 1939). I based my earliest mixes on modifications of their formulas by using soil and compost to replace the loam to which I did not have access at the time. The addition of real loam as the soil ingredient in the blocking mix recipe on page 126 could make it even better.

4. T. E. Odland, R. S. Bell, and J. B. Smith, "The Influence of Crop Plants on Those Which Follow: V," Bulletin No. 309 (Kingston, RI: Rhode Island Agricultural Experiment Station, 1950).

5. B. Gagnon and S. Berrourard, "Effects of Several Organic Fertilizers on Growth of Greenhouse Tomato Transplants," *Canadian Journal of Plant Science* 74, no. 1 (1994): 167–68.

6. Harry A. J. Hoitink and P. C. Fahy, "Basis for the Control of Soil Borne Pathogens with Compost," *Annual Review of Phytopathology* 24 (1986): 93–114; H. A. J. Hoitink, Y. Inbar, and M. J. Baehun, "Status of Compost-Amended Potting Mixes," *Plant Disease* (September 1991): 869–73.

7. A. C. Bunt, *Modern Potting Composts* (University Park: Penn State University Press, 1976).

8. Occasionally, if the mix is a shade too moist, some blocks may fall out when you lift the blocker. If you first tip the blocker slightly with a quick twisting motion before lifting it, you can break the moist suction between the soil blocks and the surface beneath them, ensuring that the blocks remain inside the blocker.

9. J. L. Townsend, "A Vacuum Multi-Point Seeder for Pots," *HortScience* 22, no. 6 (1987): 1328.

10. Redi-Heat heavy-duty propagation heat mats are sold by Johnny's Selected Seeds.

Chapter 19: Weeds

1. H. A. Roberts and Patricia A. Dawkins, "Effect of Cultivation on the Numbers of Viable Weed Seeds in Soil," *Weed Research* 7 (1967): 290–301.

Chapter 20: Pests?

1. C. E. Yarwood, "Predisposition," in *Plant Pathology*, vol. 1, edited by J. G. Horsfall and A. E. Dimond (San Diego: Academic Press, 1959), 521–62.

2. S. Perrenoud, *Potassium and Plant Health* (Worblaufen-Bern, Switzerland: International Potash Institute, 1977).

3. T. C. R. White, "The Abundance of Invertebrate Herbivores in Relation to the Availability of Nitrogen in Stressed Food Plants," *Oecologia* 63 (1984), 90–105.

4. Cited in Rene Dubos, "An Inadvertent Ecologist," *Natural History* 85, no. 3 (1976): 8–12.

5. A. H. Lees, "Insect Attack and the Internal Condition of the Plant," *Annual Biology* 13 (1926): 506–15.

6. S. H. Wittwer and L. Haseman, "Soil Nitrogen and Thrips Injury on Spinach," *Journal of Economic Entomology* 38, no. 5 (1945): 615–17.

7. F. Chaboussou, "Cultural Factors and the Resistance of Citrus Plants to Scale Insects and Mites," in *Fertilizer Use and Plant Health: Proceedings of the 12th Colloquium of the International Potash Institute* (Worblaufen-Bern, Switzerland: International Potash Institute, 1972), 259–80.

8. White, "The Abundance of Invertebrate Herbivores."

9. A. M. Primavesi, A. Primavesi, and C. Veiga, "Influences of Nutritional Balances of Paddy Rice on Resistance to Blast," *Agrochemica* 16, nos. 4–5 (1972): 459–72.

10. P. A. Van Der Laan, "The Influence of Organic Manuring on the Development of the Potato Root Eelworm, *Heterodera rostochiensis*," *Nematology* 1 (1956): 113–25.

11. Von H. Thiem, "Über Bedingungen der Massenvermehrung von Insekten," *Arbeiten über physiologische und angewandte entomologie aus Berlin Dahlem* 5, no. 3 (1938): 229–55.

12. United States Department of Agriculture, *Soil: The Yearbook of Agriculture* (Washington: USDA, 1957), 334.

13. T. W. Culliney and D. Pimental, "Ecological Effects of Organic Agricultural Practices on Insect Populations," *Agricultural Ecosystems Environment* 15 (1986): 253–66.

14. G. S. Williamson and I. H. Pearse, *Science, Synthesis and Sanity* (Edinburgh: Scottish Academic Press, 1980), 315.

15. Cited in T. B. Colwell, "Some Implications of the Ecological Revolution for the Construction of Value," in *Human Values and Natural Science*, edited by E. Laszlo and J. B. Wilbur (New York: Gordon and Breach, 1970), 246.

16. Robert Van den Bosch, *The Pesticide Conspiracy* (Garden City, NY: Doubleday, 1978), 19.

17. Meadow seed blends whose blooms provide nectar, pollen, and habitat for predatory wasps, lacewings, ladybugs, and other beneficial insects are sold under brand names such as Good Bug Blend and Border Patrol. Of course, the same plantings can also provide habitat for pest insects. Researchers have investigated a number of management strategies for enhancing the effect of the beneficials. There are scientific references in Mary Louise Flint's *Pests of the Garden and Small Farm* (Davis: University of California, 1990) that provide more information to growers interested in exploring that option.

18. Pieterse, Corne M. J. et al. "Induced Plant Responses to Microbes and Insects," *Front Plant Sci.* 2013;4:475.

Chapter 21: Pests: Temporary Palliatives

1. There are quite a number of both seaweed-based and non-seaweed-based "plant enhancing" nutrient sprays used by organic growers. An extensive selection of them is available from Peaceful Valley Farm Supply.

2. A good study with which to begin investigating this subject is Heinrich C. Weltzein, "Some Effects of Composted Organic Materials on Plant Health," *Agriculture, Ecosystems, and Environment* 27 (1989), 439–46.

   Compost Tea Manual 1.1, by Elaine Ingham and Michael Almes, is available for $15 US from Growing Solutions, Inc., 1702 W. 2nd Avenue, Eugene, OR 97402; (541) 343-8727. It provides a clear explanation of the compost tea idea as well as information on production methods and formulas for enhancing the effect of compost teas in different situations.

3. Walter Ebeling, "Sorptive Dusts for Pest Control," *Annual Review of Entomology* 16 (1971): 123–58.

Chapter 22: Harvest

1. A good (although dated) book on farmwork efficiency is *Farm Work Simplification* by L. M. Vaughan and L. S. Hardin (New York: John Wiley & Sons, 1949).

2. See *Growing for Market* magazine, January 16, 2013.

Chapter 24: Season Extension

1. Some interesting books on microclimate are: *Climates in Miniature* by T. Bedford Franklin (London: Faber, 1945); *Shelterbelts and Windbreaks* by J. M. Caborn (London: Faber, 1955); and *Climate and Agriculture* by Jen-Hu Chang (Chicago: Aldine, 1968).

2. For a manual of general information on all aspects of greenhouses, see *Greenhouse Engineering* by Robert A. Aldrich and John W. Bartok Jr. (Storrs: University of Connecticut, n.d.). Other books are listed in the bibliography.

Chapter 25: The Movable Feast

1. The only history of mobile greenhouses I have been able to find is a 1960 study done by an English garden writer with funding from a Royal Horticultural Society trust: I. G. Walls, *Design, Cropping, and Economics of Mobile Greenhouses in Britain and the Netherlands* (R. H. S. Paxton Memorial Bursary, 1960). It proved so difficult to locate that I eventually begged a copy of the manuscript from the author. Photocopies of that manuscript were recently available from ATTRA (see the sidebar in on page 227 for the address).

Chapter 26: The Winter Harvest Project

1. H. A. Roberts and Patricia A. Dawkins, "Effect of Cultivation on the Numbers of Viable Weed Seeds in Soil," *Weed Research* 7 (1967): 290–301.

Chapter 27: Livestock

1. H. R. Bird, "The Vital 10 Percent for Poultry," in United States Department of Agriculture, *Soil: The Yearbook of Agriculture* (Washington: USDA, 1948), 90–94.

2. My young layers on range have not been bothered by hawks. But if that is a problem where you live with hawks or owls, you might investigate guard dogs or guard geese or use covered pens. There are excellent designs in Joel Salatin's *Pastured Poultry Profits* (Swoope, VA: Polyface, 1993). This exceptional book also contains extensive material on feeding, slaughtering, and marketing. The author is a specialist on range poultry, and his information is very sound.

IMAGE CREDITS

The images on pages 1, 10, 13, 14, 19, 23, 31, 33, 35, 36, 38, 39, 48, 51, 63, 75, 82, 84, 85, 88, 89, 94, 100, 103, 108, 111, 113–130, 133–136, 142, 147, 151, 154, 158, 161, 164, 165, 168, 171, 173, 177–183, 187–191, 196–207, 215, 219–254, and 258; and the top image page 80, the right image on page 175, and the top image on page 217 are by **Barbara Damrosch**.

The images on pages 7 and 28 are by **Lynn Karlin**.

The photographer of the image on page 17 is unknown.

The images on pages 20 and 167 are by **Robbie George**.

The images on pages 37, 105, 184, and 209; the bottom two images on page 170; and the left image on page 175 are by **Clara Coleman**.

The image on page 41 is by **Tom Jones** for *Maine Times*.

The images on pages 66, 67, 73, 132, 137, and 194; the bottom image on page 80; and the top image on page 170 are by **Eliot Coleman**.

The images on pages 78, 110, and 143, and the bottom image on page 112 are from **Johnny's Selected Seeds**.

The image on page 83 is by **James Baigrie**.

The image on page 145 is by **Stephen Orr**.

The images on pages 148 and 195, and the top image on page 112 are by **Adam Lemieux**.

The image on page 169 is by **Jonathan Dysinger**.

The plans on page 216 and the bottom of page 217 are used courtesy of **Eric Chase Architecture**.

The map on page 255 is used courtesy of the **USDA**.

INDEX

Page numbers followed by f refer to figures. Those followed by t refer to tables.

Hoitink, Harry, 10
Holland, 200, 208, 223
hoof-and-horn meal, 126, 127
Hoop Coops, 218–20
Hopkins, Cyril G., 4, 83, 94–96
horse manure, 213–14
    in blocking mix, 125, 127
Horticultural Abstracts (journal), 225
Howard, Albert, 3, 5, 8, 9
Human Values and Natural Science
    (Colwell), 11
humidity for crop storage, 174
humus, 4, 20–21, 64
Huxley, Aldous, 231

I

Illinois System of Permanent Fertility, 94
immunity to pests, 7, 8
income per acre, 30
information resources, 221–28, 261–70
    learning from experience in, 16–17,
        226, 230–32, 236
    in libraries, 224–26
    other farmers as, 16–17, 224, 226–27
infrared-transmitting plastic mulch, 190
inoculants, bacterial, for nitrogen
    fixation, 64
insects
    beneficial, 9, 70, 154f, 159
    as pests. *See* pests
irrigation
    capital investment in, 30
    in setting out transplants, 134
    water supply for, 22–23, 25

J

Jang JP-1 seeder, 110f, 111
Japan, subscription marketing in, 181, 182
Japanese beetles, 165
Jefferson, Thomas, 6
Jerusalem artichokes, 98
Johnny's Selected Seeds, 29, 121, 147,
    150, 259
journals as information resource, 225

K

kale, 207f, 244–45
    availability by month, 44t
    in crop rotation, 56t, 57t, 58f, 59f
    green manure under, 68
    in soil blocks, 131t, 244
    soil fertility for, 91
    spacing of, 57t, 137, 139t, 244
    transplanting of, 117t, 131t, 137, 139t, 244

kaolin clay in pest control, 166
kelp meal, 90, 126
knives for harvesting, 169
kohlrabi, 45t, 218, 220
komatsuna, 207f
Krilium, 231

L

labor, 31–34
    in caterpillar tunnel ventilation, 193
    in farm stand, 176
    in harvesting, 171–72
    and importance of rest and
        reflection, 50
    in livestock care, 213
    in row cover ventilation, 191
lamb's quarters, 142f
land, 19–26
    access to, 23, 40, 43, 46, 176
    acreage of. *See* acreage
    air drainage of, 21–22
    aspect of, 21
    cost of, 23, 25–26
    in European model, 28
    for farm-generated fertility, 101–2
    geographic location of, 23
    for ideal small farm, 19, 26
    pollutants in, 25
    security of, 23–25
    slope of. *See* slope of land
    soil in. *See* soil
    subdivision of production areas in,
        43–44
    sun exposure of, 21, 22, 210–12
    water on, 21, 22–23, 25
    wind protection in, 22, 189–90, 194
Landsberger green manure mix, 69
Large, E. C., 222
Lawrence, W. J. C., 122
layout of growing areas, 43–44, 46
    length and width of beds in, 46
    subdivision of fields in, 43–44
lead in soil, 25, 230
leeks
    availability by month, 45t
    in soil blocks, 125
    spacing of, 138t, 139t
    transplanting of, 117t, 138t, 139t
legumes
    in crop rotation, 54, 56, 58, 59, 60, 69
    deep-rooted, 8, 60, 65, 98, 221
    as green manure, 8, 63, 64, 65–66,
        69–70
    nitrogen fixation by, 94

Leopold, Also, 159
lettuce, 245
    availability by month, 44t, 45t
    in crop rotation, 53–54, 56t, 57t
    as greenhouse crop, 43, 49, 198, 200,
        208–9
    harvest of, 169, 172, 174, 245
    marketing and sale of, 176, 181
    planting dates for, 49
    post-harvest treatment of, 174
    in soil blocks, 120f, 122, 124, 131t,
        208–9, 245
    soil fertility for, 91, 245
    spacing of, 57t, 122, 137, 138t, 139t, 245
    transplanting of, 49, 113, 117t, 122,
        131t, 137, 138t, 139t, 208–9, 245
    winter production of, 43, 208–9
ley farming, 9, 106–7
Ley Farming (Stapledon), 106
library resources, 224–26
Life Reform movement, 3
lifestyle, as motivation for farming, 18
lime and limestone, 4, 8, 14
    application rate, 85, 95
    in blocking mix, 125–26
    in crop rotation cycle, 55
    on forage fields, 101
    in maintenance of soil fertility, 107
    and pH, 85, 90, 95, 125–26
    tillage of, 73
livestock, 213–20
    acreage for, 25
    cattle, 107, 213
    chickens, 214–20. *See also* chickens
    and crops in mixed farming, 8, 106–7,
        214, 218–20, 233
    green manure as feed for, 65, 70
    horses, 213–14
    manure from, 213–14. *See also* manure
Living the Good Life (Nearing), 42
long-handled hoes, 144–45
low plant covers for season extension,
    190–92
lupines as green manure, 65, 74

M

mâche, 45t
major crops, 42, 43
    availability by month, 44t
    list of, 42t
management of employees, 33–34
Man and the Soil (Mickey), 5
manure, 8, 83–84
    application rate, 84, 95, 213–14

ABOUT THE AUTHOR

Barbara Damrosch

Eliot Coleman has over fifty years of experience in all aspects of organic farming, including field vegetables, greenhouse vegetables, rotational grazing of cattle and sheep, and range poultry. He is also the author of *Four-Season Harvest* and *The Winter Harvest Handbook*, as well as the instructional workshop DVD *Year-Round Vegetable Production with Eliot Coleman*. Coleman and his wife, Barbara Damrosch, presently operate a commercial year-round market garden, in addition to horticultural research projects, at Four Season Farm in Harborside, Maine.